设计DESIGN 源自生活 设计创新生活
FROM LIFE INNOVATION LIFE

AN ESSAY ON DESIGN
CHAN MAN TUNG

陈文栋设计随笔

CHAN TAN FUNG　HUANG CUI PING
EDITED　GRAPHIC DESIGNED BY
黄翠萍　陈丹枫
编辑　美术设计

FU XING　CHEN WEI XIONG
WROTE　PHOTOGRAPHED BY
陈伟雄　傅兴
摄影

CHAN MAN TUNG
WROTE　ILLUSTRATED BY
陈文栋
文　插图

CWD DESIGNERS (HK) LTD
PLANNED BY
香港陈文栋设计师有限公司
策划

LIAONING SCIENCE AND TECHNOLOGY PUBLISHING HOUSE
SHENYANG
辽宁科学技术出版社
沈阳

图书在版编目（CIP）数据

陈文栋设计随笔/陈文栋著. —— 沈阳：辽宁科学技术出版社 2016.2
ISBN 978-7-5381-9540-8

Ⅰ.①陈… Ⅱ.①陈… Ⅲ.①建筑设计－文集 Ⅳ.
①TU2-53

中国版本图书馆CIP数据核字(2016)第000461号

◆

出版发行：辽宁科学技术出版社
地　　址：沈阳市和平区十一纬路29号 邮编：110003
印　刷　者：恒美印务（广州）有限公司
经　销　者：各地新华书店
幅面尺寸：230 mm × 305mm
印　　张：30.5
插　　页：5
字　　数：400千字
出版时间：2016年2月第1版
印刷时间：2016年2月第1次印刷
责任编辑：于　倩
封面设计：黄翠萍
版式设计：黄翠萍
责任校对：李淑敏
书　　号：978-7-5381-9540-8
定　　价：258.00元

◆

联系电话：024—23284356　13130216749
邮购热线：024—23284502
E-mail:yuqian_mail@126.com
http://www.lnkj.com.cn

豪华尽享

才能回归平淡
奢侈的尽头，便是简单、自然。

RETURN TO FLAT AFTER ENJOYING ALL
IT IS SIMPLE AND NATURAL AT THE END OF THE LUXURIES

中国房地产

让豪宅半成品交楼，成为历史的设计大师

"新古典"精装设计掌门人

让中国新富阶层，居住品质与世界同步的思想者

让土豪更豪，让平凡不凡的达人

著名职业经理人，优秀的建筑师

成功的室内设计大师

◆

China Real Estate
To deliver the half-finished luxury real estate, to become the design master of history
Head of the designer for "New Classics" fine decoration
A thinker who provides a synchronous world living quality with the new rich in China
A expert to make the rich richer, let the ordinary extraordinary
Famous professional managers, excellent architect
and successful master of interior design

中国·广东·华南理工大学78级　建筑学学士
South China University of Tech. (1978)　Bachelor of Arch.

国际室内装饰设计协会　资深会员/常务理事
IFDA　Senior fellow and general council

1991—2001	祈福集团 Clifford Group	设计总监 Design Director
2001–2002	合生创展 Hopson Group	设计总监 Design Director
2002–2013	星河湾集团 Star River Group	设计总监/副总裁 Design Director / Vice President

香港陈文栋设计师有限公司　设计总监/创始人
CWD designers (hk) Ltd.　Design Director / Founder

开篇语

春有百花秋望月，夏有凉风冬听雪。

心中若无烦恼事，便是人生好时节。

在这人生的好时节，重新整理廿多年职业生涯值得留恋的设计作品，

算是一件幸事。

往惜的：

宝贵光阴，青春年华，艰苦岁月，辛劳磨难，已是流水般逝去……

眼前的作品带给你的，尽是：

喜悦、欣慰、回忆、享受……

这是付出的奖赏，耕耘的收获。

骐骥一跃，不能十步。

驽马十驾，功在不舍。

这是三十年前自勉的座右铭。

将耳顺之年了，再次用优美的作品，愉悦你的双眼，温暖我的心房。

INTRODUCTORY WORDS

Spring flowers and autumn moon, summer breeze and winter snow.

If there is no trouble in mind, it is a good time for life.

In the good time of life, it is a blessing to rearrange memorable designs in my career that is more than twenty years.

In the past:

Precious time, youth, hard times, tribulations all passed like flowing water...

What the present work brings you are:

Joy, relief, memories, enjoyment...

Those are rewards for pay, harvest for cultivation.

Steed leap, not ten steps, ten inferior horse riding, Gong-in give up.

This was my motto thirty years ago.

I am almost sixty years old, but again I would please your eyes and warm my heart with beautiful works.

让平凡不凡

Poltrona Frau、意大利全皮家具，100年历史，皇家御用。
简约、时尚、色彩缤纷。
法国85岁大画家Hemenet，精彩作品，私人订制。
一望无际的风景，美不胜收的原野，
深邃的海洋，灿烂的罂粟花，
奔放的大色块，浪漫的小情怀，
天才的创造力，惊艳的想象力。
蔚蓝的天空，让窗户灵魂出窍。
金黄的石材，让土豪更加土豪。
变幻的天花，让平凡不再平凡。
温馨、绚丽、精致、大气的氛围，
让"新古典"风格，
再次展现了崭新、亮丽的画面。

Let Ordinary Extraordinary

Italian leather furniture Poltrona Frau with a history of 100 years is a supplier for the Royal.
Simple, fashion and colorful.
Wonderful works of 85-year-old French painter Hemenet, personal tailor.
Endless landscape and breathtaking wilderness,
Deep ocean and brilliant poppies,
Unrestrained big color lumps and romantic feelings,
Genius creativity and amazing imagination.
Blue sky has freed the soul of window,
Golden stone makes the rich richer.
Changeable ceilings let the ordinary extraordinary.
Cozy, gorgeous, delicate and generous atmosphere.
"New Classic" style unfolds a brand-new and bright picture.

经典简约 时尚大宅
CLASSIC SIMPLE   FASHION HOUSE

积跬步 至千里

不积跬步，无以至千里。
不积小流，无以成江河。

设计随笔，积十几年涓涓小流。
辛勤劳动、努力耕耘、点点滴滴，
愿汇成一首动听的歌、一幅美丽的图画，
给人以片刻的欣赏或享受，
亦让自己偶尔陶醉其中。

设计工作是枯燥无味的，像刻板的"四平拳"，需要磨炼与浸淫。
但"艰难困苦，玉汝于成"。
设计工作是丰富多彩的，像美妙的古巴"莎莎"，天赋纵情地发挥，"舞姿"散发魔鬼般的魅力。
感动你周边的人，也感动你自己。

A short step to a thousand miles

A short step to a thousand miles
Rome was not built in one day.

Design essay records more than ten years of designs bit by bit.
Wish to make a pleasant song and a beautiful picture with hardworking and every little things.
Provide people with moment appreciation and enjoyment, also draw myself in it occasionally.

Design work is boring just like rigid "siping fist" which needs to be cultivated and immersed.
However, "difficulties and hardships help to sing".

Design work is rich and colorful just like splendid Cuba Salsa, displaying talent like no one's watching,
and the dance exudes the charm of the devil.
Touch people around you and yourself.

设计与时尚

设计追求时尚,因为时尚是美。
设计必须创新,时尚就是创新。
美是百看不厌,可惜世上没有永恒。
昨日的灿烂辉煌,今天已是夕阳黄昏;
昔日的超级明星,今天已失去光芒;
过去的青春动人,今天已岁月留痕。长江后浪推前浪!
时尚就像那长江中滔滔不尽的后浪。
她推动着社会的文明进步!
令世界色彩缤纷,让人们焕然一新。
追求时尚吧,但切忌走火入魔。

Design and Fashion

Design pursues fashion because fashion is beauty.
Design must be innovative because fashion is an innovation.
Beauty is something that we never get tired of;
however there is no eternity in this world.
Yesterday's brilliant abs became today's dusk.
Former superstar has lost their shine.
Past youth has gone old. The waves pushed before.
Fashion is the endless waves behind in the Yangtze River,
she drags the progress of social civilization!
Add colors to this world and fresh people.
Pursue fashion but not be obsessed.

色彩感觉

色彩因配搭得宜而产生美，如同蓝天衬白云，红花映绿叶，才子择佳人，宝剑配"霸王"。

色彩如多变的风景：

红热情如火，蓝冷若冰霜，绿如迷人的远山，紫如天边的云雾，

黄似丰收的田野，黑像深邃的大海，白是飘舞的雪花……

人们喜爱不同的色彩，如同钟情不同的季节。

高大有型的，喜欢冬天的长裘、大衣；青春动人的，迷恋夏天的短裙、背心；

成熟稳重的，爱上春天的绿地、高球；一见钟情的，思念秋季的枫叶、小道。

色彩喜好因彼此的感觉相异而不同，

人们随着心境的变化、年龄的增加、阅历的丰富、见识的提高，

感觉与喜好会移情别恋，见异思迁，朝秦暮楚……

喜爱不是一成不变的，变 —— 才是永恒！

Sense of Color

Colors generate beauty because of suitable match, like blue sky and white clouds,

red flowers and green leaves, gifted scholars and beautiful ladies,

Color is like changeable scenery: red is warm like fire, blue is cold as frost,

green is charming as mountain, purple is like clouds of the sky, yellow is the harvest field,

black is the deep ocean and white is the flying snow...

People have the same love feelings for colors and ocean.

The tall like the long overcoat in winter, the young are infatuated with shirt and vest in summer;

mature people love green land of spring,

people fall in love at the first sight miss the maple and trail in autumn.

Color preferences are different from each other because of different feelings.

People change their preferences along with the change of mood, age, experience.

Love is not static, change is the only eternity!

❋ 再论风格 ❋

"风格"这个话题烦死人,"风格"这个话题你必须面对,否则更烦。

风格是什么?

风格就是一种文化,一种内涵,一种艺术修养,一种独特创意的表达。

不同历史、环境、地域、区域,产生着不同的文化与创意,

百花齐放,争妍斗丽,能脱颖而出者成为"风格"。

并非所有的作品都具有称"风格"的资格,

只有优秀的代表性的作品才能佩戴"风格"的头衔。

设计师并非为风格而创作,

风格只是一种符号,风格也可是一代宗师。

但设计师追求的是文化的提升,修养的锤炼,学识的发展。

设计的目的,是解决难题,赢得市场,

将各种门派会聚一体,创一套"截拳道",你就是李小龙。

什么风格,由人说去。

"把风格抛之脑后,将创意发挥,让梦想飞驰,

作品也就脱胎换骨,耳目一新,精彩万分……"

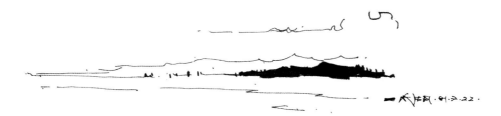

About Style

"Style" is an annoying topic, but when "style" is the topic you have to face, it becomes more annoying.

What is style?

Style is a culture, a connotation, an artistic rhetoric and a unique expression of innovation.

Different history, environment, regional and area produce different culture and originality.

People who stand out from the crowd can be the "style".

Not all works has the qualification to be called as style.

Only the outstanding representative works can wear the title of style.

Designers do not create for style, style is a symbol or is a great master.

Designers shall pursuit the promotion of culture,

accomplishment of temper and development of knowledge.

The goal of design is to solve tough problem and win the market.

If you can create a set of "Jeet kune do" through a comprehensive study of the subjects then you are Bruce Lee.

No matter what people say, be creative and fly your dreams by leaving the style behind,

then you can have creative and fresh work.

设 计 与 构 想

没有专业构思的过程，室内装修就只有装修，没有设计。
何为构思？
构思是建立在专业知识基础上的想象力，创造力。
没有这一基础，你也可以构思，如同没有文化，你也可以写作，结果不言而喻，不堪设想。
室内设计学，包含了艺术的属性，它需要丰富的想象力，崭新的创造力。

Design and Ideas

Interior renovation only has decoration when there is no process of professional conception.//
What is conception?
Conception is imagination and creativity based on professional knowledge.
Without this foundation,
you can still design just like you can write without knowledge,
but the outcome is obvious.
Interior design includes the nature of art,
it needs rich imagination and brand new creativity.

❖ 设 计 与 灵 感 ❖

世上没有"灵感"。
灵感,是长期积累、艰苦创作、反复推敲、柳暗花明时遇上的一个可爱的村庄。
只要你是有心人,"灵感"是"踏破铁鞋无觅处,得来全不费工夫"。
因此,不要责备自己缺乏天才与灵感,关键是你是否已"踏破铁鞋"而到达"山穷水尽"的境地,
从来没有经历这种艰辛与困境,自然就得不到"灵感"的眷顾。
"灵感"如同树上的花朵,需要不断地培育,否则它会枯萎。
"灵感"如同甜蜜的"爱情",梦想一劳永逸,不可能"天长地久"。

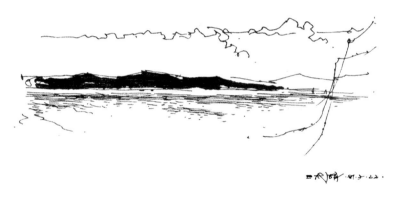

Design and Inspiration

There is no "inspiration" in this world.
Inspiration needs long-term accumulation, hard work and repeated scrutiny.
As long as you are a man of heart,
inspiration will show just like you wear out iron shoes in hunting round,
must come all not to time-consuming.
Thus, don't blame yourself for lack of talent or inspiration,
the key is that if you have worn out your iron shoes and reached the wits'end.
People won't have inspiration unless you have been through all these.
"Inspiration" is the flower on the tree which needs to be cultured constantly, otherwise it will wilt.
"Inspiration" is like sweet "love" dreaming to be forever when there is no eternity.

创新风格　各自精彩

五星级酒店，1200 平方米总统套房。
将新古典发挥得淋漓尽致。
让人目不暇接，惊喜万分。
喜欢就是喜欢。
倘若不能打动你，
只因曾经豪华享尽。

设计的奥妙在于：
不断的创新与变化。
风格虽各自精彩，
追求的共同目标，始终是美。
学海无涯，创意无限。
扬长避短，就是唯一。

Innovative Style and Individual Beauty

Five star hotel with 1200 square meter presidential suite.
Exceed the new classic to the fullest.
Give people the feelings of dizzying and surprising.
Like be like, if cannot touch you, the only reason is that you have enjoyed all the luxury.

The secret of the design is continual change and innovation.
Different styles have different shining points, but the common goal is always beauty.
Learning is endless and creativity is infinity.
The only way is to enhance advantage and avoid disadvantage.

让土豪更豪　让平凡不凡
MAKE THE RICH RICHER AND ORDINARY EXTRAORDINARY

价值观

体现人的价值，物的价值，是要投奔"怒海"，这个"怒海"，就是"市场"！

什么是好的作品？

不是"欧式"，不是"法式"，更不是"古典式"，也不是"现代式"。

形式、风格并不重要。

设计真正"出彩"才重要，说"出彩"，很抽象，"出彩"就是市场的接受与认同，

"出彩"如同一首深入人心的流行曲，它唤起人们的共鸣，深入人们的文化，体现时代的精神与品味。

作品让市场来检验，很残酷也很公平，

不同的地区与国家，不同的时间与年代，市场的标准各不相同，

因此，设计师应该是一个"变色龙"，一个"百变高手"。

Values

Entering the market can embody human values!

What is a piece of good work?

It is not "European style", "French style", "classic style" or "modern style".

Form and style are not important.

The real important part is the shining point of a design. Shining point is very abstract to say,

it is the acceptance and recognition of the market,

"shining point" like a popular song can evoke sympathy and embody the spirit of the time.

It is cruel and fair to let the market test.

Different areas, countries, time and ages have different market criteria,

thus designers should be a "chameleon" or a "variety master".

经典与创新

审美的重复会令视觉疲劳,
设计中要有惊喜,不断创新就成为必然。
如果"经典"能代表永恒,
那么时尚就是刹那的光芒。
让经典再现,让作品时尚,设计才会发亮、发光……

Classic and Innovation

Visual fatigue comes from repeated aesthetic.
Ongoing innovation is necessary for surprise in design.
If "classic" can present eternity,
then fashion is the moment light.
Design can only shine if the classic can reappear and work is fashionable.

时代的宠儿

始于2005年，
一支"土八路"设计团队，
不畏国际大牌，不畏压力重重，
打造了"时代的宠儿"，
享誉京城，满载而归。
成为学习的典范，超越的对象。
实在是：
有心栽花花不开，无心插柳柳成荫。

非常感谢这个时代，
欣赏、宠爱"新古典"的人，
"新古典"的铁杆粉丝……

The Darling of the times

The native design team was set up in 2005,
with no fear of international brands and pressure,
We created "the darling of the times",
became well-known in Beijing and win a lot of honors.
We became the learning model and the object to be gone beyond.
It was a coincidence.

Thanks to this era,
and the person who enjoy and love "new classic",
the fans of "new classic"…

"新古典"精装 时代的宠儿
DELUXE BINDING OF "NEOCLASSIC" DARLING OF THE TIMES

硬装与软装

何谓"硬装"?
"硬装"就是室内空间的设计,是你的形,你的五官与肢体。
何谓"软装"?
"软装"就是室内家私、灯饰的布置与一切摆设,是穿衣打扮,涂脂抹粉。
"帅哥",给他一个时尚的发型,得体的衣着,
潇洒的风度将自然流露,挥洒自如。
耀目的领带,闪亮的皮扣,
乌黑的皮鞋,心仪的腕表。
一切细节都展现着你的品味和内涵,不容有失!
"美眉",要神彩飞扬,同样要靠那殷红的双唇,飘逸的秀发,碎花诱人的短裙。
"软装"充分展现人的个性与风格,
"硬装"却是一切梦想,"神彩飞扬"、"颠倒众生"的金钥匙。
"硬装"是一个表演的舞台,
"软装"是舞蹈家、歌唱家、钢琴大师、功夫之王,相互辉映、相得益彰,
完整的室内设计,原来包括了"硬装"与"软装",缺一不可。

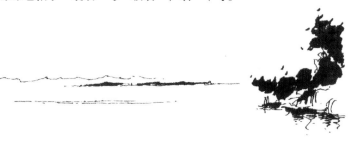

Interior Architecture and Soft Decoration

What is Interior architecture?
Interior architecture is the design of interior space, like your shape, the facial features and body.
What is Soft decoration?
Soft decoration is the decorative units such as furniture, lamps, like the dress and make up.
Handsome man can be charm of his manner by giving him a stylish haircut and decent clothes.
Shiny tie and leather buckle, dark black leather shoes and favorite watch,
every detail can revel your taste and connotation.
Girls need red lips, nice hair and shirt to be brimmed with health and spirits.
"Soft decoration" can show people's personality and style, but "Interior architecture" is every dream,
 and the key to be in high spirit and make your charms irresistible.
"Interior architecture" is a stage, and "Soft decoration" is dancers,
singers, piano masters, kung fu champion, they bright out the best in each other.
The complete interior design includes both "interior architecture" and "soft decoration".

设计的灵魂

室内设计,"光"是她的灵魂,
没有了光,空间就没有了生命。
不同形式的光,令空间产生不同的气氛。
"光"可以营造明快、舒适、热烈、温馨、浪漫、宁静等气氛,
也可展现丰富多彩、诡秘或者风和日丽的情景。
因此,不要吝啬光的存在,它带给你的只会是愉悦与钟爱,
满足与享受。

Soul of the Design

Light is the soul of interior design, without light, there is no life in space.
Different forms of light create different atmosphere.
Light can create sprightly, comfort, enthusiastic, cozy,
romantic and peaceful atmosphere,
also can show a scene of colorful, mysterious and sunny.
Thus, don't stingy light, it can bring you pleasure, love,
satisfaction and enjoyment.

室内之美

设计必须创新，时尚就是创新。
室内设计的美，包含了各种精彩的细节，
如绘画般的构图、似音乐般的作曲、像诗人般的填词、绚丽、优美、动人……
丰富多彩的室内空间，像美妙的古巴"莎莎"，
天赋纵情地发挥，"舞姿"散发着魔鬼般的魅力。
感动你周边的人，也感动你自己！

Indoor beauty

Design must have innovation, and fashion is the innovation.
The beauty of interior design includes all kinds of gorgeous, graceful and moving details...
Colorful interior space is like wonderful Cuba Salsa which displays talent like no one's watching,
and the dance exudes the charm of the devil.
Touch people around you and yourself.

品质的追求

温馨、绚丽、大气、夺目、奢华、耀眼,从来都不是刻意追求,
高品味的设计,高品质的营造,
让一切自然而然、尽情流露……
愉悦你的双眼,温暖你的心窝。
关于美 —— 所有艺术追求的共同目标。
美 —— 就是打动你,就是眼前一亮,就是让视觉完全地满足!
美 —— 完整性、恰当的比例、漂亮的色彩,缺一不可。

Pursuit of quality

Cozy, colorful, majestic, sparkling,
luxurious and dazzling are never the deliberate pursuit.
Design with good taste and creation with high – quality make everything look natural.
Please your eyes and warm your heart.
Beauty – the common goal of all arts.
Beauty – is to touch you, to brighten you and fully satisfy your vision.
Beauty – is integrity, appropriate proportion and beautiful color,
none is dispensable.

风格与作品

设计风格、流派，
均是在作品诞生之后被人们有意无意戴上的桂冠。
设计风格直接地反映着设计师内在的文化、艺术、知识、修养与潜在的创意。
设计师不是在为风格而创造作品，
但作品却在无声无息地显现作者的风格，
因作品是在不断创新与完善的，
因此"风格"也就不是一成不变的，它只是一种代名词。
"设计师应该把所谓风格抛诸脑后，
不要被它所捆缚，将创意发挥，让想象飞驰，
作品也就可以脱胎换骨，风格也就耳目一新，精彩万分。"

Style and Works

Design styles and genres both are crowns intentionally or
unintentionally put on by people after the birth of great works.
Design style reflects the internal culture, art, knowledge,
cultivation and potential originality.
Designer doesn't create works for style, but works can revel designer's style silently.
Works are innovated and improved constantly,
thus style is not static but one kind of synonym.
Designer should leave the so called style behind,
do not be tied by it and free your imagination.
Works and style can be fresh and wonderful.

关于美

美感——是一切艺术形式追求的共同目标。

美是完整性的代名词，

是比例恰当的写照，

是美丽色彩的化身。

室内设计所追求的是像音乐一样的旋律美、节奏美，像绘画一样的视觉美、色彩美；

像雕塑一样的形态美、立体美；像诗歌一样的意识美、感受美。

美感是在统一中求变化，在变化中求统一。

没有统一的变化，似是乱弹的琴；

没有变化的统一，像沉闷的歌。

完整性包括了统一、协调、和谐、均衡、对称……而美，

只需一句话来概括，就是当她展现在人们面前时，

"让人获得完全的满足"！

In regard of Beauty

Aesthetic feeling is the common goal pursued by all kinds of arts.

Beauty is the synonym of integrity and the incarnation of beautiful colors.

Interior decoration pursues rhythm beauty as music, version beauty as pictures,

shape beauty and three-dimensional beauty as sculpture and awareness beauty as poetry.

Beauty pursues changes in unification and unification in changes;

there is no unified changes and varying unification.

Integrity includes unification, coordination, uniformity and symmetry.

Use one sentence to summarize,

it would be "give people complete satisfaction" when she was displayed in front of people.

也谈艺术

何谓艺术？艺术是客观世界与主观意识的碰撞，是客观现实的存在性与主观世界的创造性相结合的产物。

"若言琴上有琴声，放在匣中何不鸣？若言声在指头上，何不于君指上听"。

琴：客观世界；指：主观意识的体现。

只有在"心"的指引下用"指"去拨动琴弦，才能产生动听的旋律与声音。

艺术所表现的是来自现实的东西，但又并非现实的原貌，它是现实的提炼与再造。

"吾师心、心师目、目师华山"。

绘画的对象是华山，但纸上表现的只是心中的华山，而并非目中的华山。

徐悲鸿的马、齐白石的虾、莫奈的莲与毕加索的裸女，都是创造的结晶。

没有创作的过程，没有提炼与再造，也就没有了艺术。

About Art

What is art? Art is the crash of objective world and subjective consciousness,

product of combining existence of objective reality and creativity of subjective world.

"If sound comes from lyre, then why there is no sound when lyre is in the box?

If sound comes from fingers, then why not listen music on your fingers".

Lyre: objective world; finger: subjective consciousness of object.

There will be beautiful melody when pluck the string with fingers under the guidance of heart.

What art express is from reality but not the original appearance of reality.

It is the refining and recycling of reality.

"Give play to subjective initiative, describe according to one's own subjective feelings".

Painting object is mountain, but the mountain on the paper is the mountain in heart not the real mountain.

Xu Beihong's horse, Qi Baishi's shrimp,

Monet's lotus and Picasso's naked woman are all the result of creation.

There is no art without process of creation, refining and reconstruction.

空间才是主题

室内设计，
不仅仅是平、立面及天花的设计，
室内设计是营造由这些平、立面及天花所组成的"室内空间"的设计。
空间才是主题，是"主角"，
其他只是它的一部分，是"配角"，令"主角"演绎得精彩，
设计才有价值，才会成功。
否则，抓不住"主题"，
即使落笔千言，也是言不达意的。

Space is the theme

Indoor design is not only the "interior space" design including plan design,
elevation design and ceiling.
Space is the theme and "protagonist",
others are just part of it,
they are "supporting roles" to help "protagonist" perform better,
this kind of design is valuable and will succeed.
Otherwise, it is unable to speak even with thousands words.

精彩与正途

理论毕竟是抽象的、理性的认识，始终是不断的感性认识，最终产生飞跃的过程。

所以说要做武林高手，需要艰苦的锻炼与长时间的浸淫。

但学习艺术，并不仅仅是学习理论，直接的感觉才最重要，感觉需要理论，需要推敲，更需要天分！

何谓天分？强烈的爱好与触电般的感觉。

如果你的孩子对于美会渴求、会陶醉、会向往、会想拥有，就让他学习艺术吧。

像毕加索、罗丹、贝聿铭、张艺谋、马友友、李云迪、郎朗……

做个让人敬仰的大画家、大雕塑家、大建筑师、大导演、大提琴家、大钢琴家……光宗耀祖，名留青史。

名留青史并不重要，活得开心并且精彩丰富，才最实在。

上帝往往喜欢捉弄人，在赐予你才华与天赋时，也带给你痛苦与磨难。

如果没拥有艺术的天分，也就不必走那艰苦的路了。

扬长避短，才是正道！

千万不要误入歧途。

Splendidness and Right track

Theory is abstract, rational knowledge, is a leap process of continuous perceptual knowledge.

Therefore martial arts master needs tough and longtime exercise.

However, learning art is not only about theory. The important part is the direct feeling.

Feeling needs theory, scrutiny and talent. What is talent? Passion and electric shock-like feeling.

If you kid yearn for, is intoxicated with, looks forward to, wants to have beauty, then let him learn art.

Like Picasso, Rodin, Ieoh Ming Pei, Yimou Zhang, Yoyo Ma, Yundi Li and Lang Lang...

Becoming a famous painter, sculptor, architects, director,

cellist and pianist to uphold the family honor and be remembered.

It is not important to be remembered or not, most important is to live a happy and wonderful life.

God likes playing with people, when he gives you talent and gift, he gives your pain and tribulations.

If you do not have talent, there is no need to choose the tough road.

It is the right way to enhance advantage and avoid disadvantage!

Don't be misguided.

成败与命运

室内设计,
建筑空间的美,决定了作品的成败。

你若有范爷的脸,丫鬟终会变凤凰,
你若有楠的高音,歌手总会夺第一。

天分、本质很重要,
但舞台更重要。
设计师的命运,就是努力赢得舞台的过程。
只有赢得舞台,
才能插上翅膀,翱翔四方……

Success or Failure and Destiny

In interior design,
beauty of architectural space decides the success or failure of works.

If you have beautiful face, you will be a queen finally.
If you have a nice high pitch, you will be the first in the game.

Talent and essence are important,
but stage is more important.
The destiny of a designer is the process of winning stage.
Only win the stage,
you can fly to everywhere…

空间决定美
SPACE DECIDES BEAUTY

也谈房地产

房地产设计，

单体设计（户型）是核心。

室内设计（精装）是临门一脚。

如同汽车产业，比拼外形的吸引与配置的精良，但核心还是"发动机"！

如同世界杯，球星云集，攻城掠地，门前混战，但没有"外星人"罗纳尔多的门前一脚，

一切都将前功尽弃、功亏一篑。

因此优秀的单体、优秀的室内设计，其价值就是车中的奔驰、宝马，

球星中的"外星人"，价值连城、妙不可言！

About Real Estate

In regard of housing design,

the core is unit design (house model).

Interior design is the last push.

As in the automobile industry,

compare to attractive appearance and fine equipment, the core still is "engine"!

As the world cup, there are many famous football stars fighting in front of the gate,

but they would not win the game without Ronaldo's final shot.

Thus good units and interior designer are valuable!

无奈的成功

建筑设计，一直伴我前行。
但室内设计的成功，却超越了我的本科。
十分无奈，亦自得其乐。
中国房地产，
早已是国际设计大师争妍斗丽的舞台，
崭新、优秀的建筑设计作品，
比比皆是。
唯有室内精装设计，
并非优秀便可胜出。
除了品质与内涵，核心是文化的理解，
与心灵的沟通。
否则，
设计的得失，市场的认同，
差之毫厘，失之千里。

Helpless Success

I have been living under the companion of architectural design for a long time.
The success of interior surpassed my own major;
I am helpless but happy as well.
Chinese real estate has become the stage for international designers,
brand new and fine architectural design works are everywhere.
Only in interior hardcover design, excellence doesn't mean you can win.
Expert quality and connotation,
the core are the understanding of culture and spiritual communication.
Otherwise:
In the pros and cons of design and market acceptance,
a least bit of difference can cause totally different result.

大气磅礴　唯有建筑
MAJESTIC ONLY EXISTS IN ARCHITECTURE

构思与表现力

设计的开始,首先是"构思",

讨论、阅读、参观、视察、了解、争论、发现、决定……

就是一个构思的艰苦而又快乐的旅程,

不露声色,冥思苦想,天马行空,芝麻开门……

构思终于完成,作品其实已经诞生,

接下来的工作是去把它表现出来。

表现力很重要,

表现力就是想象力!

表现力就是创造力!

如果,构思无法表现出来,那么不要谈你的想象力、创造力,

因为一切都是空谈!

为什么当今的大导演、大设计师,

都是一个绘画的高手。

因为绘画就是表现力,

因为绘画是一切美学入门最基本的途径。

Conception and Expression

Conception is the start of design, then there are discussion,

reading, visit, inspection, comprehension, argument, discovery and decision...

Conception is a process full of hardness and happiness.

After a long time effort, conception is produced and the work is finished.

The next part is to express it.

Expression is very important, expression is imagination and creativity.

If the conception cannot be expressed,

then forget about your imagination and creativity,

because everything is moonshine.

Why do the great director and designer have the ability to draw,

because drawing is expression.

Drawing is the most basic way to know about aesthetics.

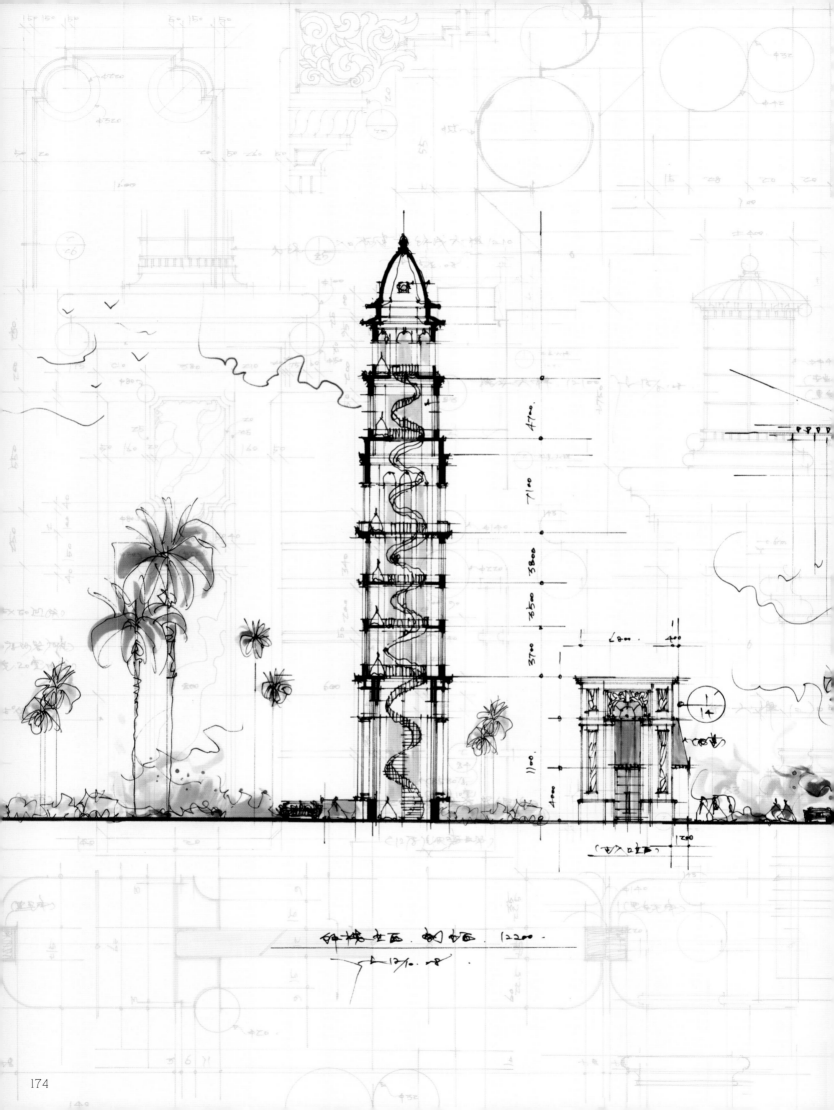

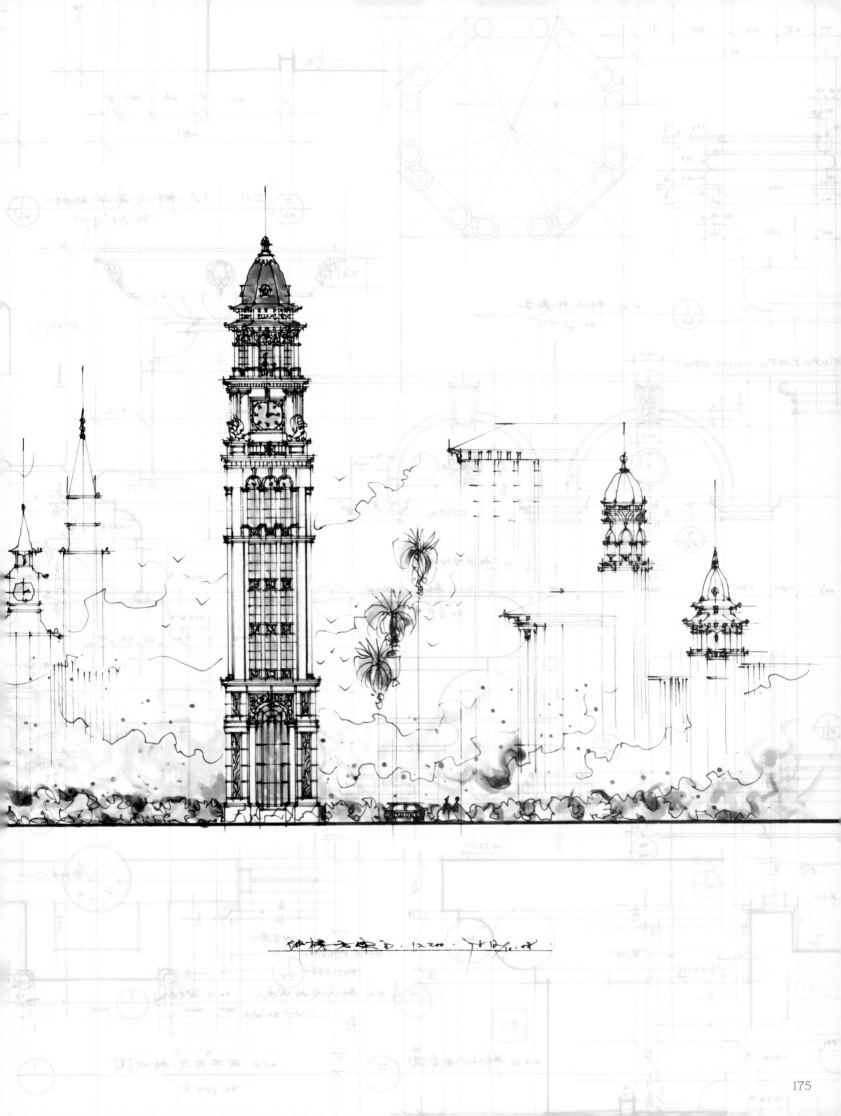

建筑的魅力

优秀的建筑，就是凝固的音乐，就是不朽的雕塑。

一座座，高低起伏，抑扬顿挫。一幢幢，鳞次栉比，井然有序。

挺拔的塔楼，仪态万千，优雅的飘檐，层层叠叠，

伸展的斗拱，俊秀的立柱，

深啡色的瓦顶，浅啡色的外墙，千家阳台，万家灯火。

裸露在蓝天白云之上，荡漾在七彩晚霞之中，大气而磅礴，高贵且亲切。

无须金雕玉砌，更胜金雕玉砌……

美 —— 所有艺术追求的共同目标。

Charm of Architecture

Excellent building is caky music and monumental sculpture.
The ups and downs buildings stand in good order.
Tall towers, elegant eaves, stretched brackets, dark coffee colored roof,
light coffee colored wall and thousands of balconies and lights.
No luxuriant decoration is better than florid decoration…
Beauty – common goals of all the art pursuit.

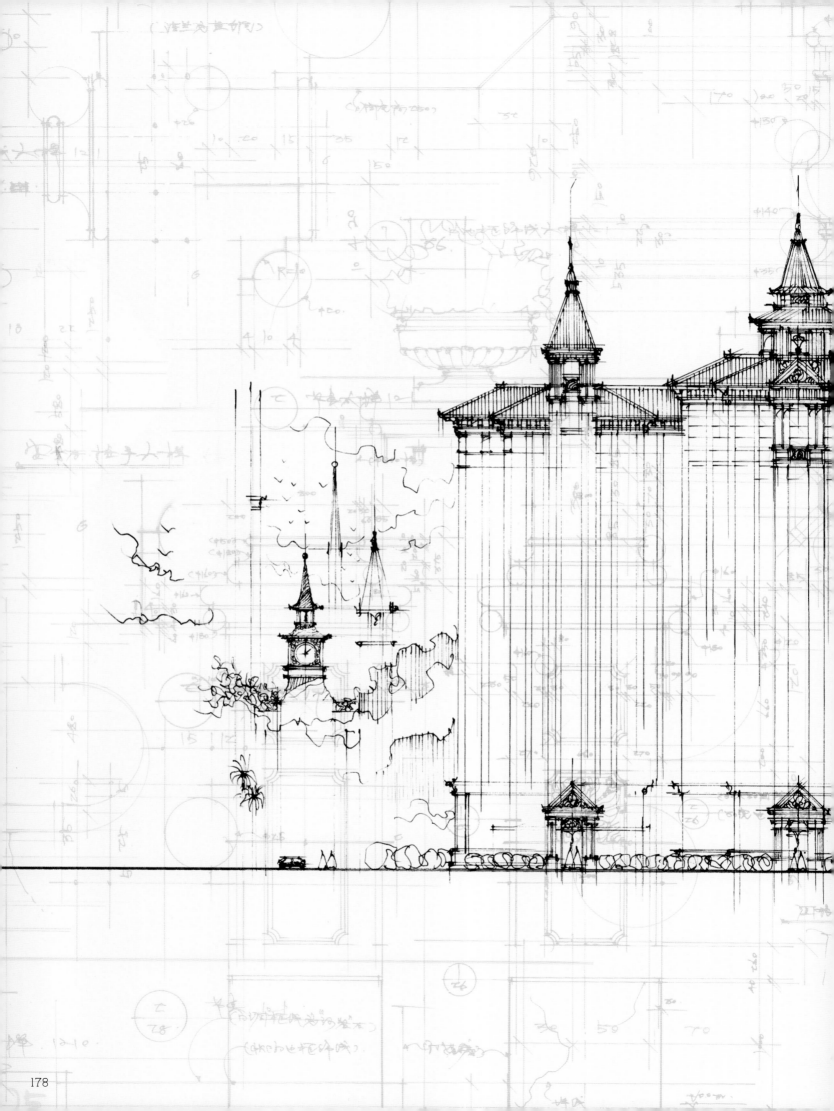

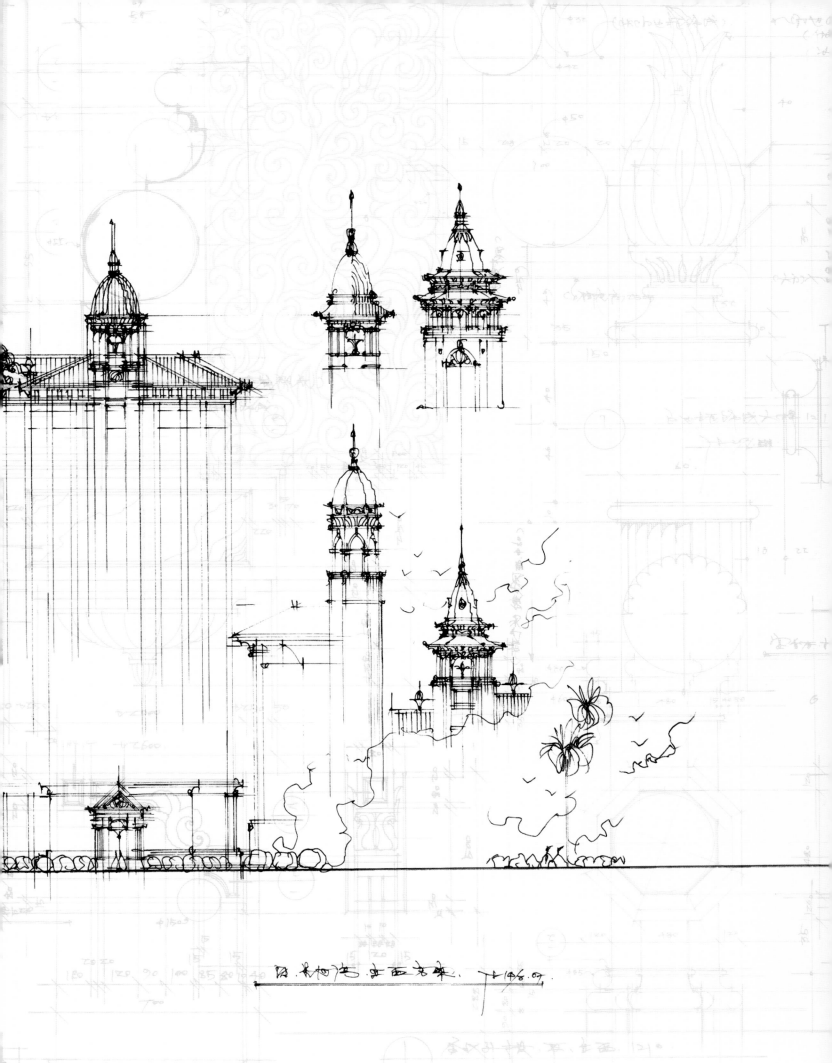

建筑设计与室内设计

建筑设计与室内设计是一对孪生兄弟，
建筑设计展开的同时，室内空间就随即产生。
"建筑师"所创造的是建筑功能的合理布局，
与建筑形态的艺术创新。
好的建筑师，
应该懂得及研究室内设计，
因为建筑形体及其比例决定了室内空间的优劣，
优秀的室内设计，
建立在优秀的建筑空间里。

Architecture and Interior Design

Architect and interior designers are a pair of twin brothers.
Interior space is produced at once when architecture design starts.
The creation of "architect" is the rational distribution of building function
and artistic innovation of building type.
A good architect should know and study interior design,
because physique of the building and its proportion decide the good and bad of interior space.
A successful interior design is built in the architectural space.

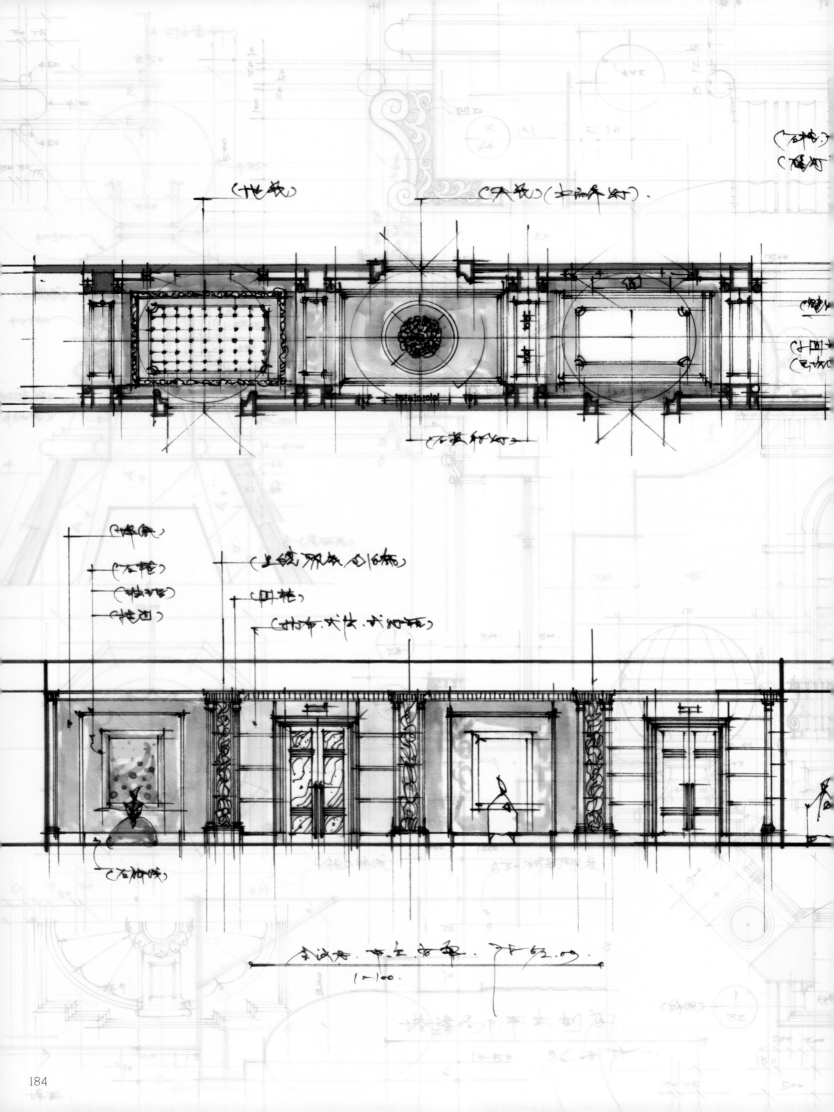

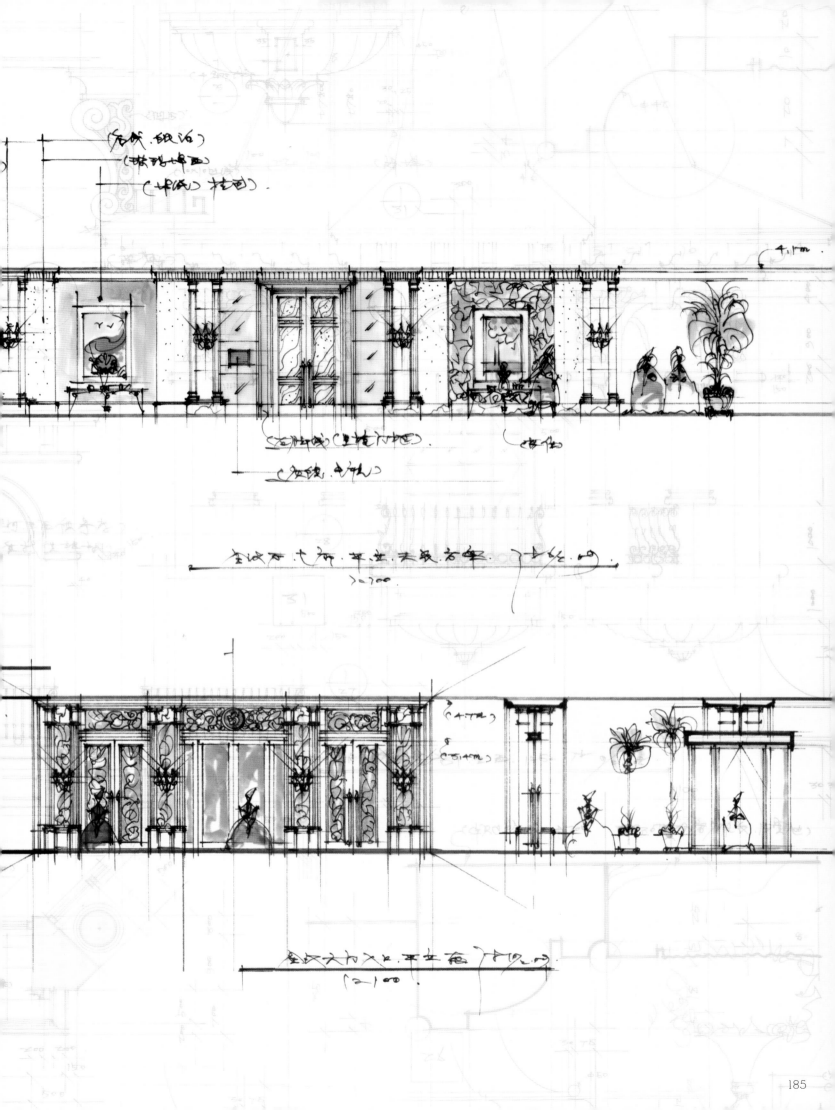

视觉魅力

视觉，撩动人的喜怒哀乐，挑逗人的食欲、性欲，

激发人的创意与冲动，唤起人的想象与憧憬，震慑人的心灵与感受……

面对优美的风景，因为舒怀与陶醉，

你可独对它，从晨光初现，直至徐徐黄昏。

面对浪漫的情景，因为身心的融化，香浓的咖啡，你可以意犹未尽，细细品尝。

面对梦中的美人，因为诱惑与迷失，你会语无伦次，身不由己、举止失措。

面对奢侈的"名牌"，因为向往与崇尚，你会一掷千金，面不改容，又或发奋工作，誓死拥有……

一个优秀的室内设计作品，她所创造的空间、气氛、色彩……

不就是一道亮丽的风景，一个浪漫的梦中情人，一件精彩、诱人的"路易威登"吗？

Vision Charms

Vision can provoke people's motion, appetite and sexual desire,

stimulate people's originality and impulsion,

arouse imagination and longings and shock people's heart and mind...

When in face of beautiful scenery,

you can face it alone from morning to dusk because of the easiness and intoxication.

When in face of romantic scenery,

you can enjoy and taste it slowly because of melting body and a cup of good coffee.

When in face of dreaming beauty,

you will push the panic button because of temptation.

When in face of luxury brands,

you will spend lots of money with calm because of longing and admiration.

A good interior design work creates space, atmosphere, colors...

Not just beautiful scenery, a romantic dream lover or a piece of attractive "LV".

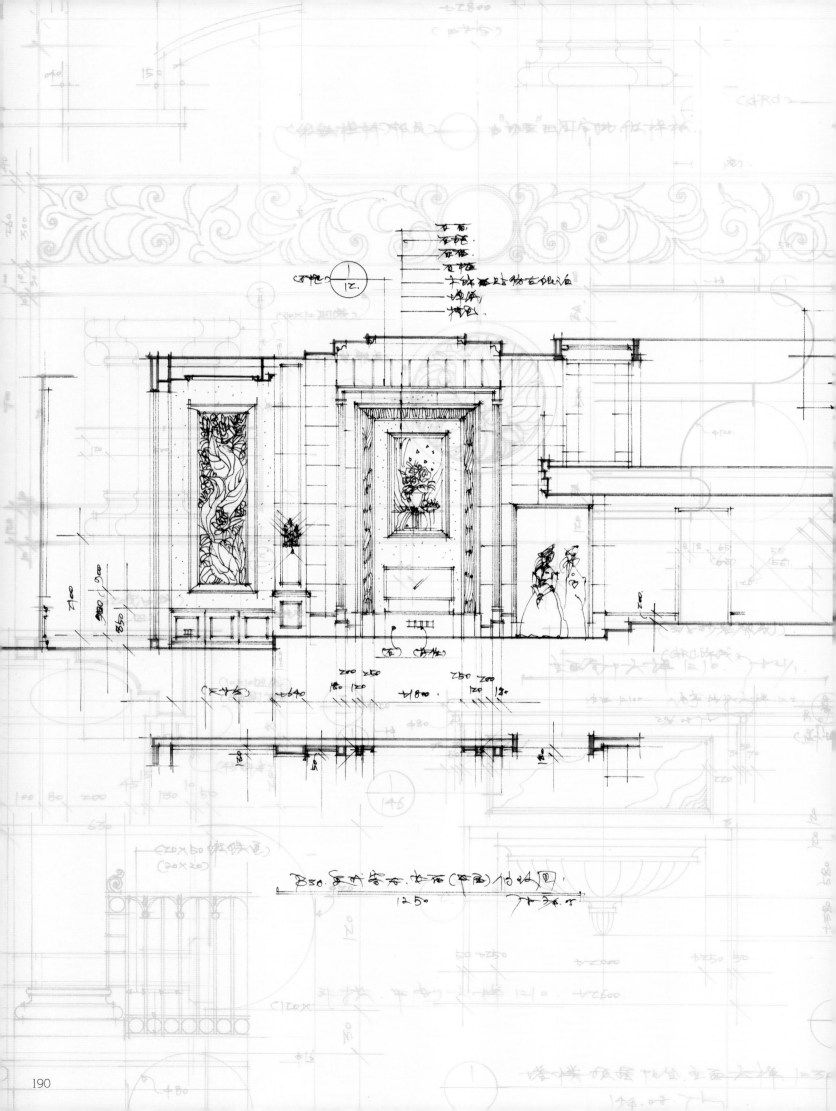

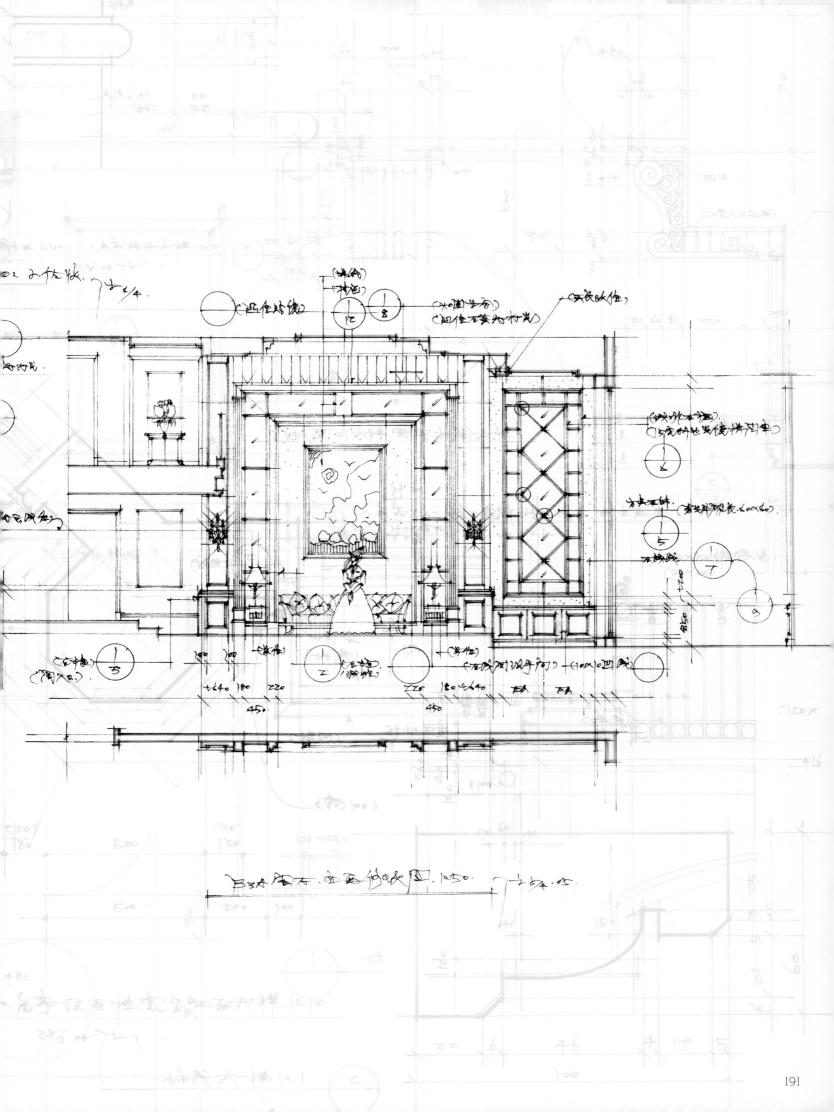

矛盾与统一

善与恶、真与伪、
美与丑、黑与白、
高与低、肥与瘦，
因对立而存在，
因比较而有鉴别。
建筑或室内设计，
同样是在矛盾中求统一，在比较中获得肯定。
设计过程是一个否定、再否定的过程，
你不敢颠覆原有的构思，
就无法创作出新的作品，
方案的比较与颠覆是直截了当、行之有效的，
是选择好作品的方式与手段，
运用这一手段，迈向成功就有了途径。

Contradiction and Unity

Good and evil, true and false, beautiful and ugly, black and white,
high and low, fat and slim.
They exist because of contradiction,
they have difference because of comparison.
Seek unification in the contradiction, find affirmation during comparison.
Process of design is the process of denial and redenail.
If you do not have the gut to overthrow the original idea,
you won't create new works.
It is a fast, effective and good way to choose a better works.
It is the path to success.

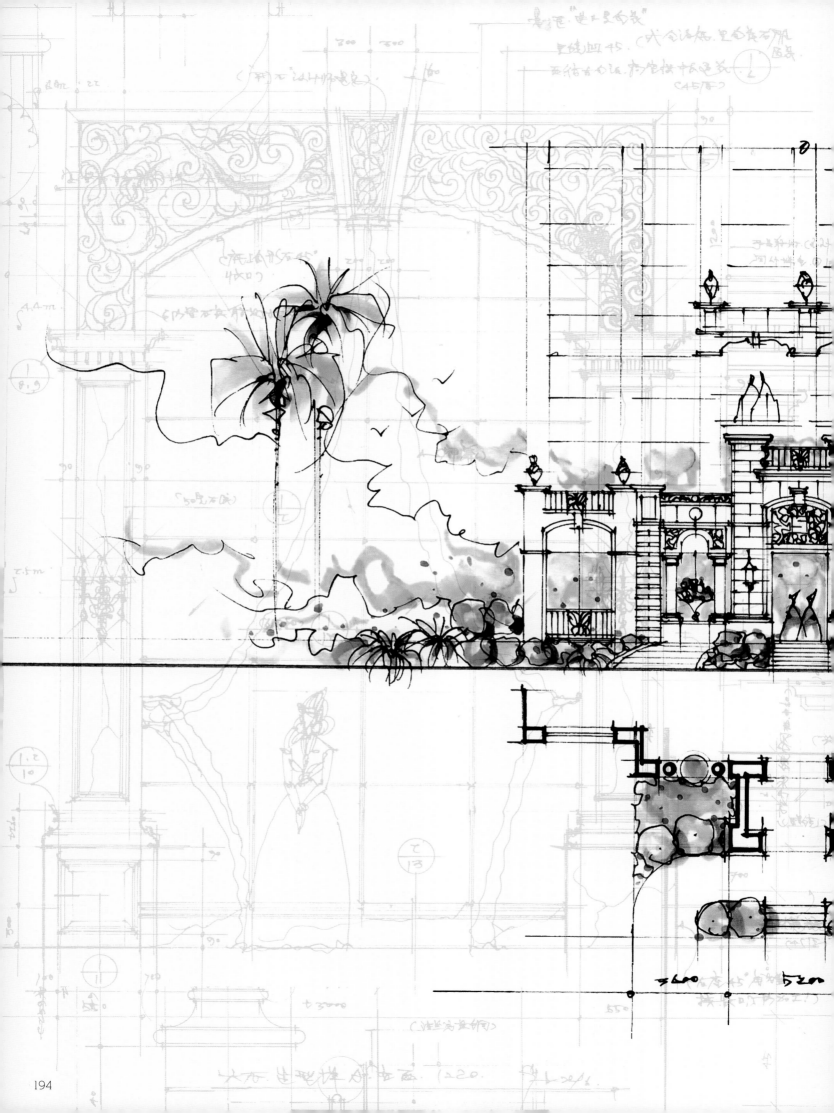

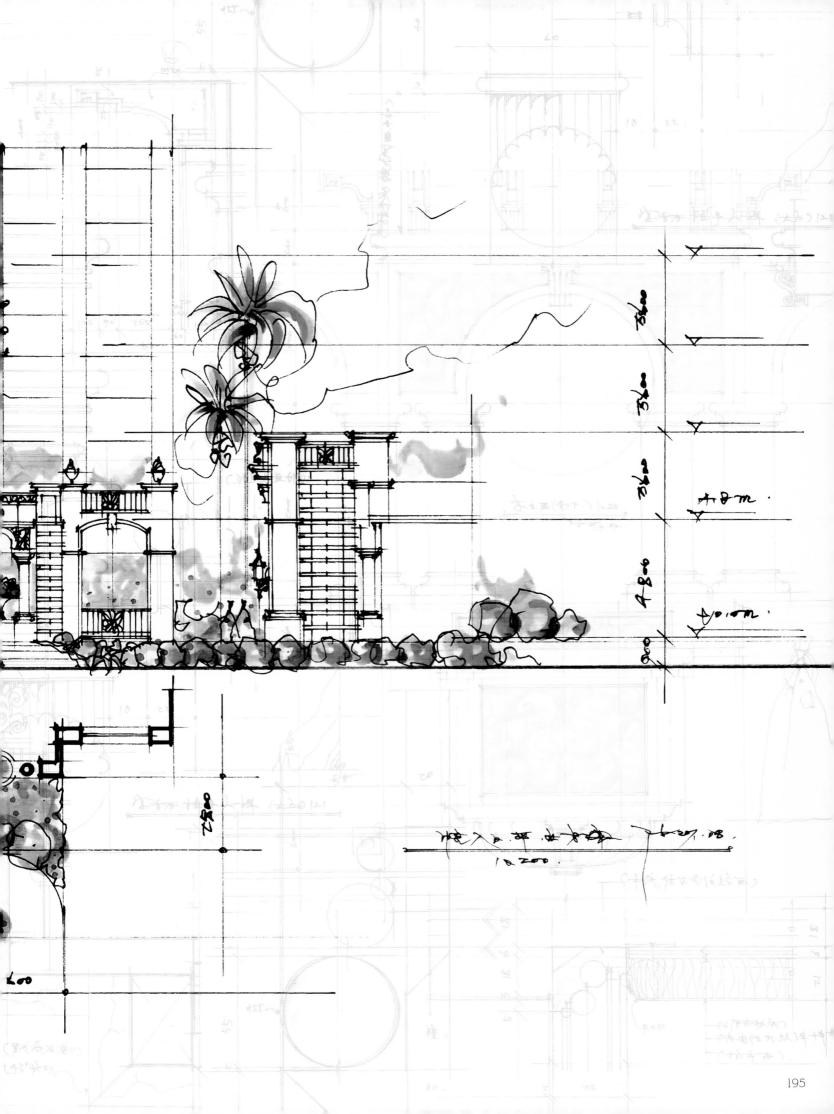

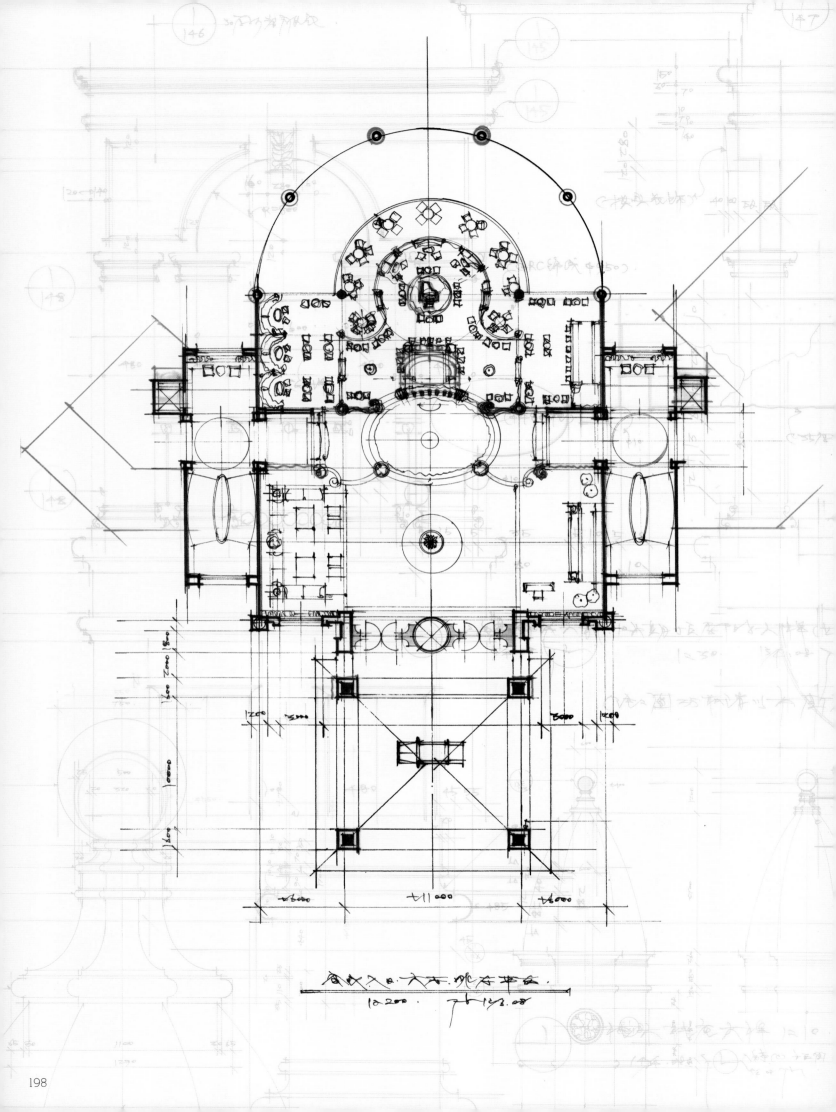

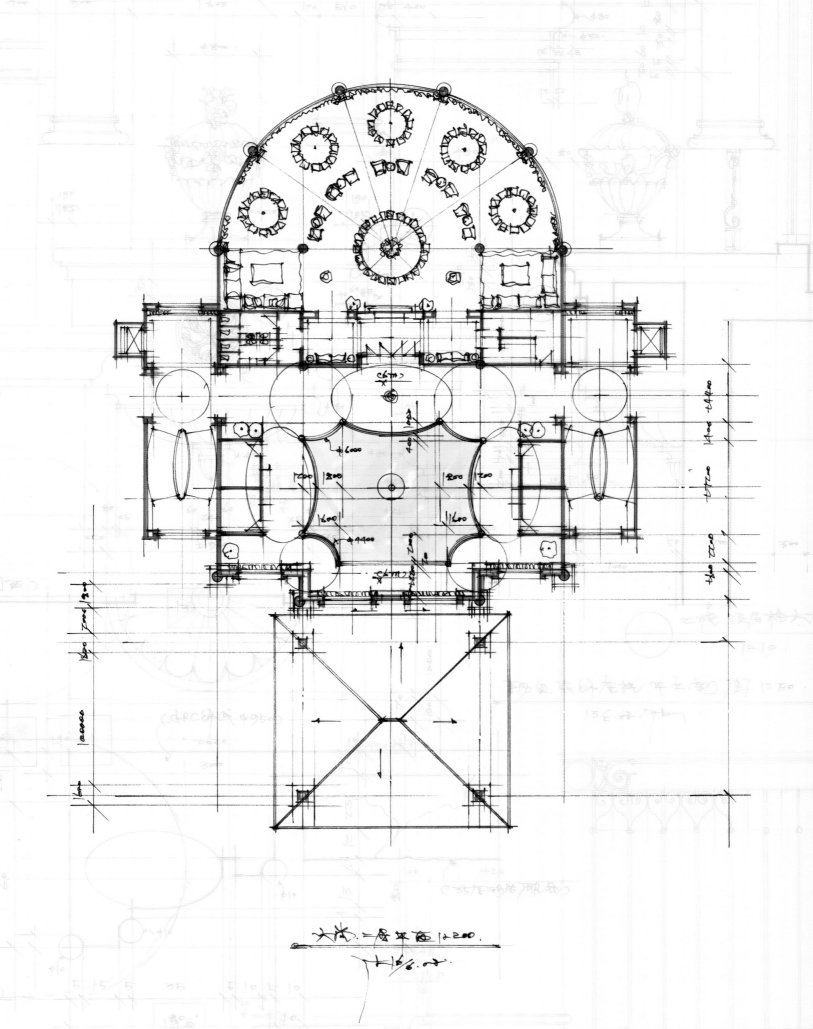

功到自然成

一代大侠江海天，与名门少侠金逐流，联手对付一位"蒙面客"，
三人激斗，各展其技，龙虎相争，险象环生，令人震惊的是"蒙面客"总占上风。
但他所施展的并非什么绝世武功，而是武林中人个个都懂得的武术中最基本的拳法"四平拳"……
上述是梁羽生的《侠骨丹心》中的一小段插曲，
当然"蒙面客"是少侠的父亲，大侠的师傅。
这里我要说的不是武功，而是室内设计。
我们的设计，就是要打"四平拳"，把基本的拳学好了、打好了、精通了、
炉火纯青了、出神入化了，你就是一个"蒙面客"。
不懂"四平拳"的人，急着创作、创新的人，都只能是"花拳绣腿"，空有其表，滥竽充数。
厚积才能薄发，滴水可以穿石，
设计中的"四平拳"是美学中最基本的原则的掌握。
完整性、恰当的比例，漂亮的色彩，心领神会自然如鱼得水，如虎添翼，要做"大侠"，功到自然成。

Effort will lead to Success

Hero Haitian Jiang and noble hero Zhuiliu Jin deal with a masked man together.
The masked man can prevail during the fierce battle.
He didn't perform any unique kung fu but the most basic fist called "Siping fist"…
That is a small episode in Yusheng Liang's "A Chivalrous Heart".
What i am talking here is not kung fu but interior design.
Our design just like playing "Siping fist", we have to learn and play well the must basic first.
If can you play it masterly, then you are the masked man.
People who don't know "Siping fist" but are eager to create,
and innovate can only hold a post without qualification.
Constant drop water wears away a stone.
"Siping fist" in interior design is the basic principle of aesthetics.
Understand integrity, appropriate proportion and beautiful color can help you a lot.

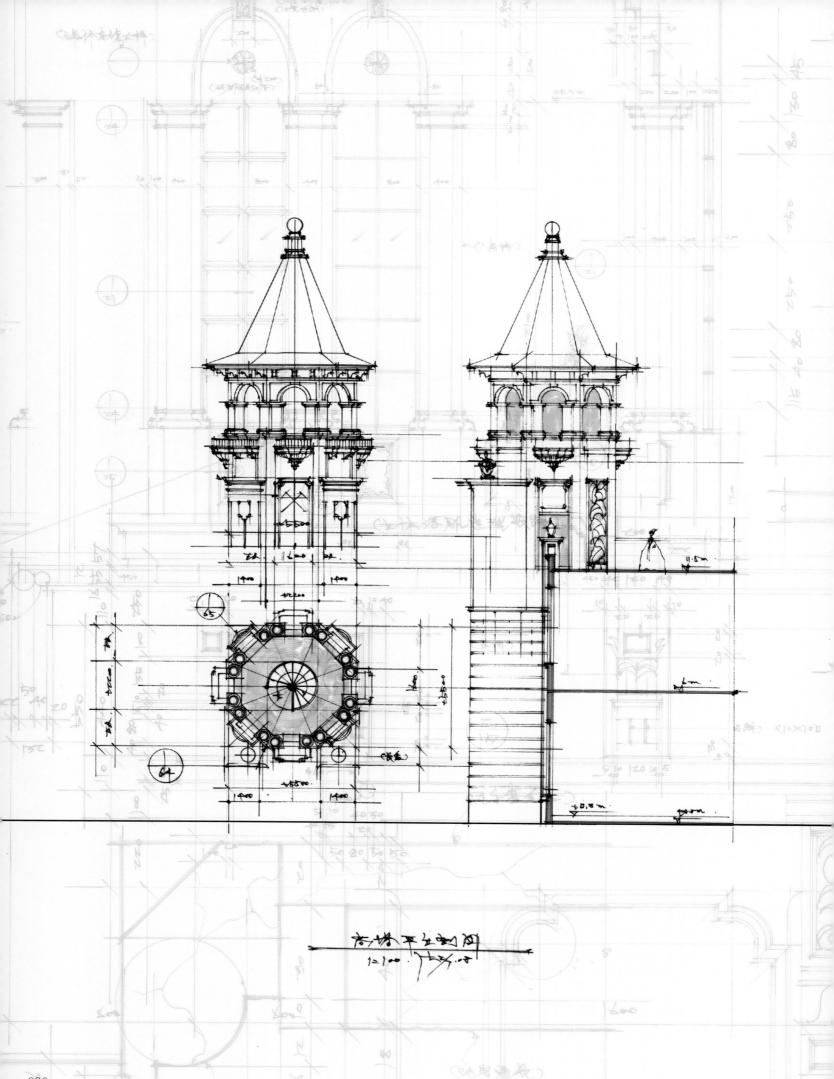

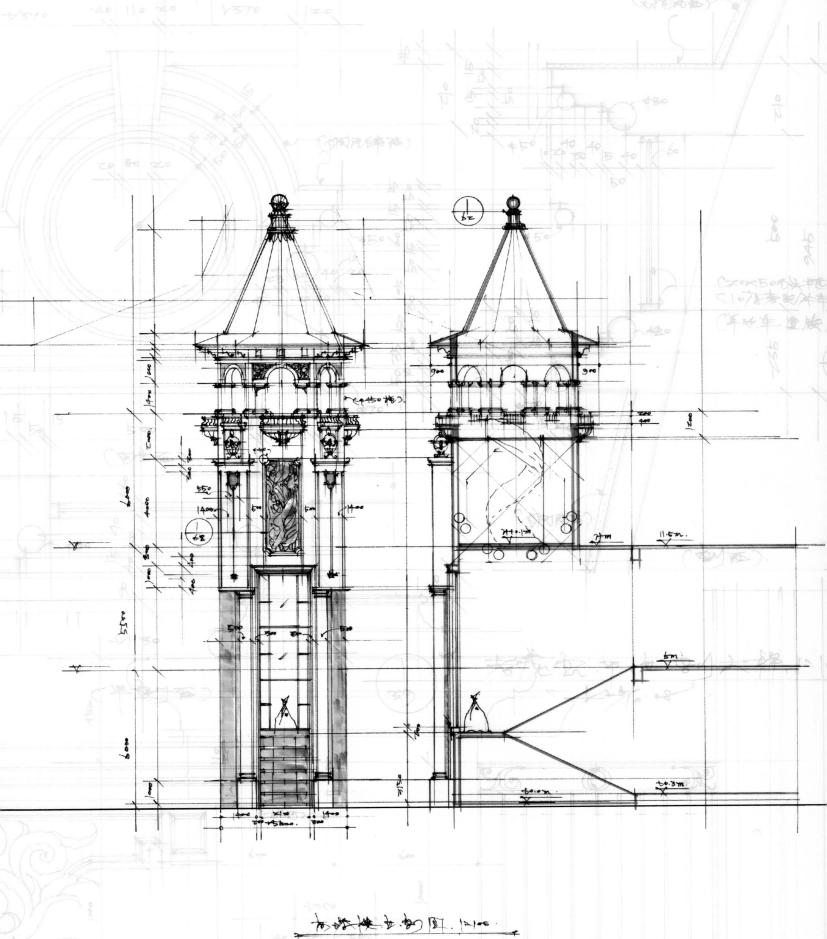

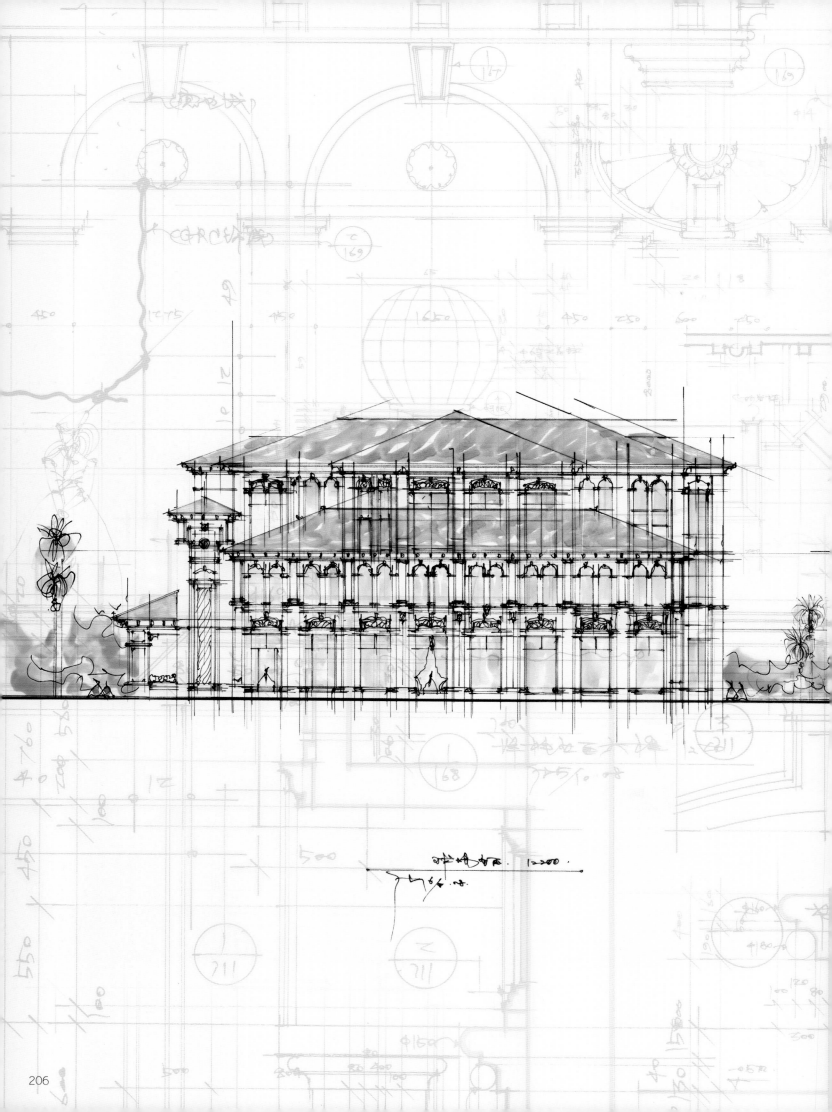

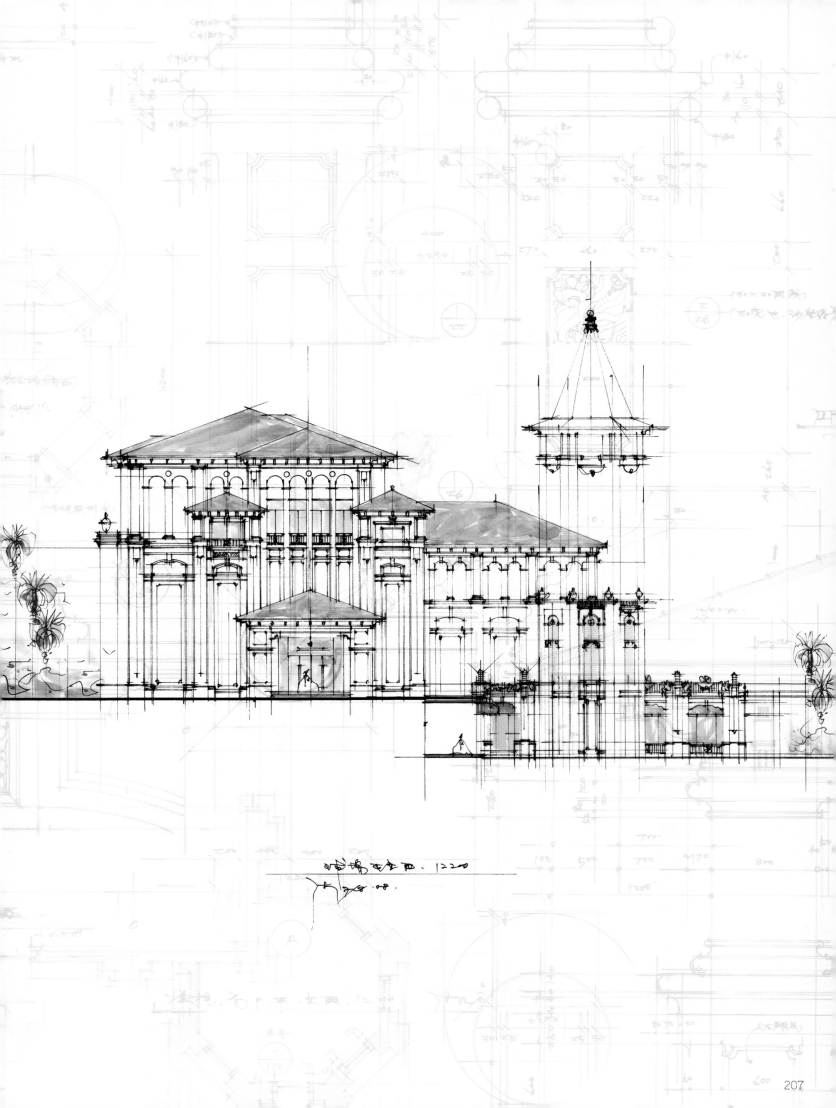

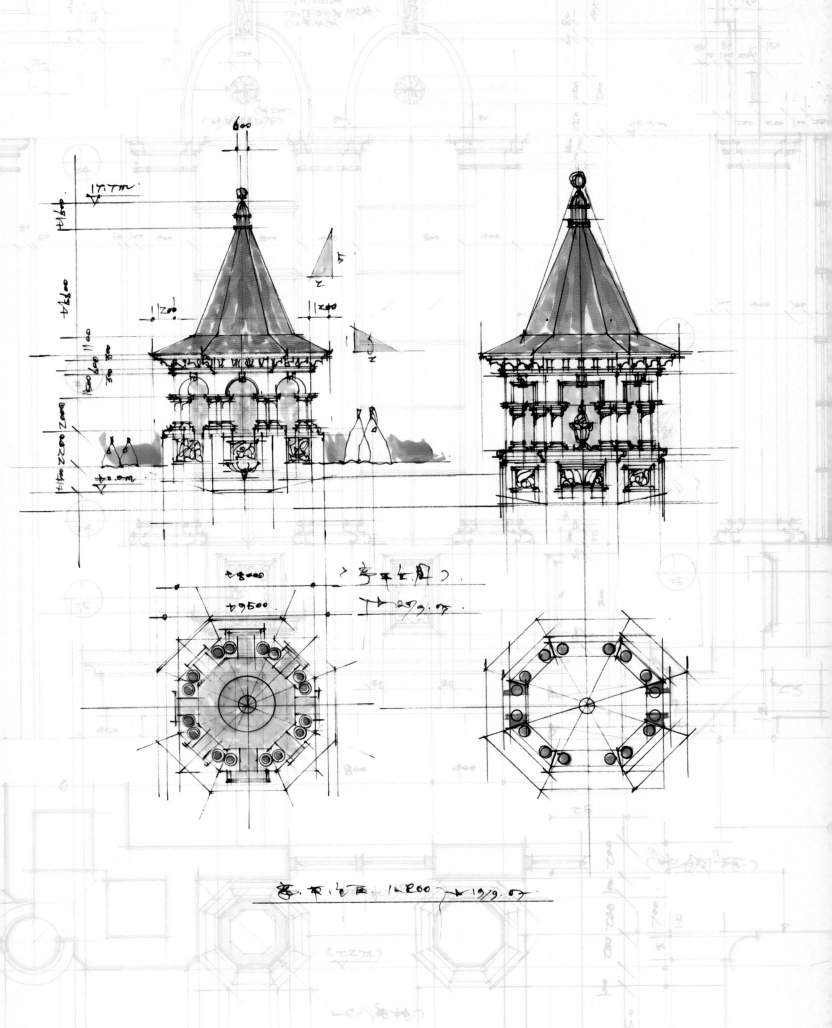

简约不简单

"简约"的精彩在于它的无拘无束，轻松自然。
我们不要说"无就是有"，"少就是多"。
我们要说的是统一与和谐，恰当的比例与漂亮的色彩。
简单的一条线，一种色，一个型，
都是经过反复的推敲、对比、选择，恰到好处，惜墨如金。
将复杂变为简单，将丰富变为平凡。
这是生活的选择，生活的升华……

Brief is not Simple

Splendidness of "simple" is free and easy.
We don't say "without is with", "less is more".
We are talking about unification and harmony,
appropriate ration and beautiful color,
Every liner, color and type are chosen after repeated consideration,
comparison and selection.
To make complex become simple and plentiful become ordinary.
These are choices and sublimation of life…

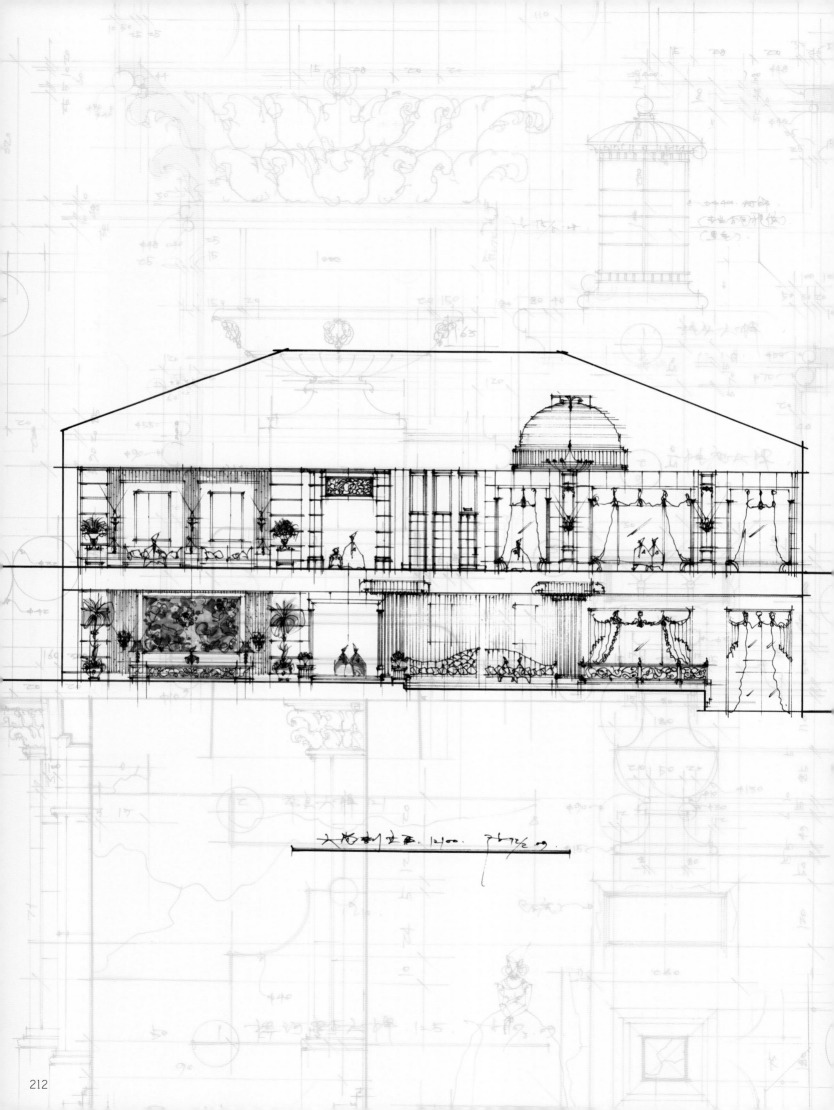

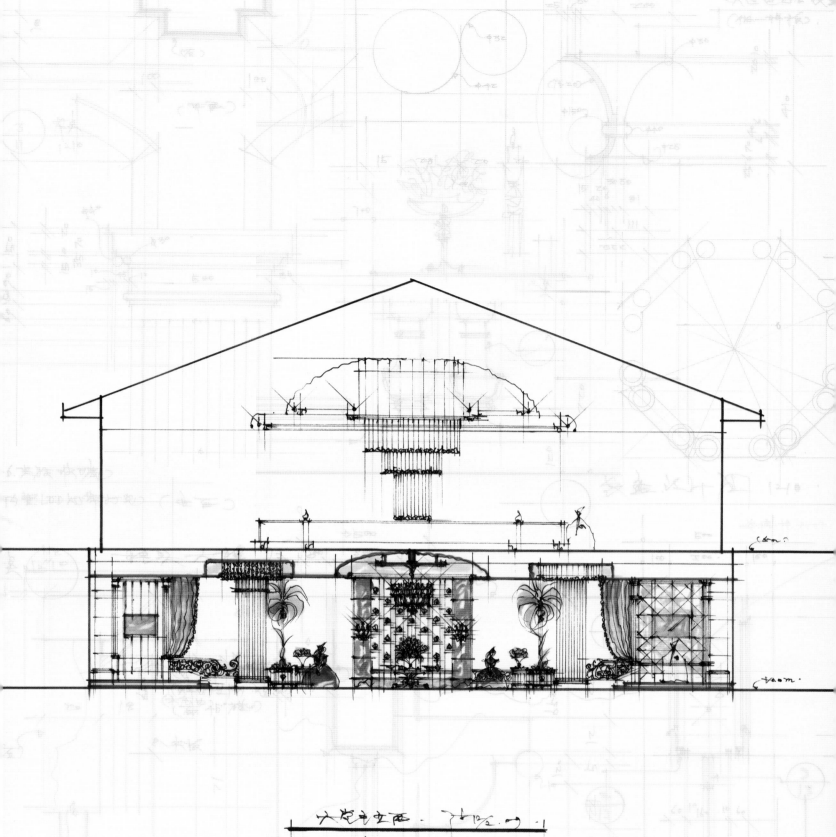

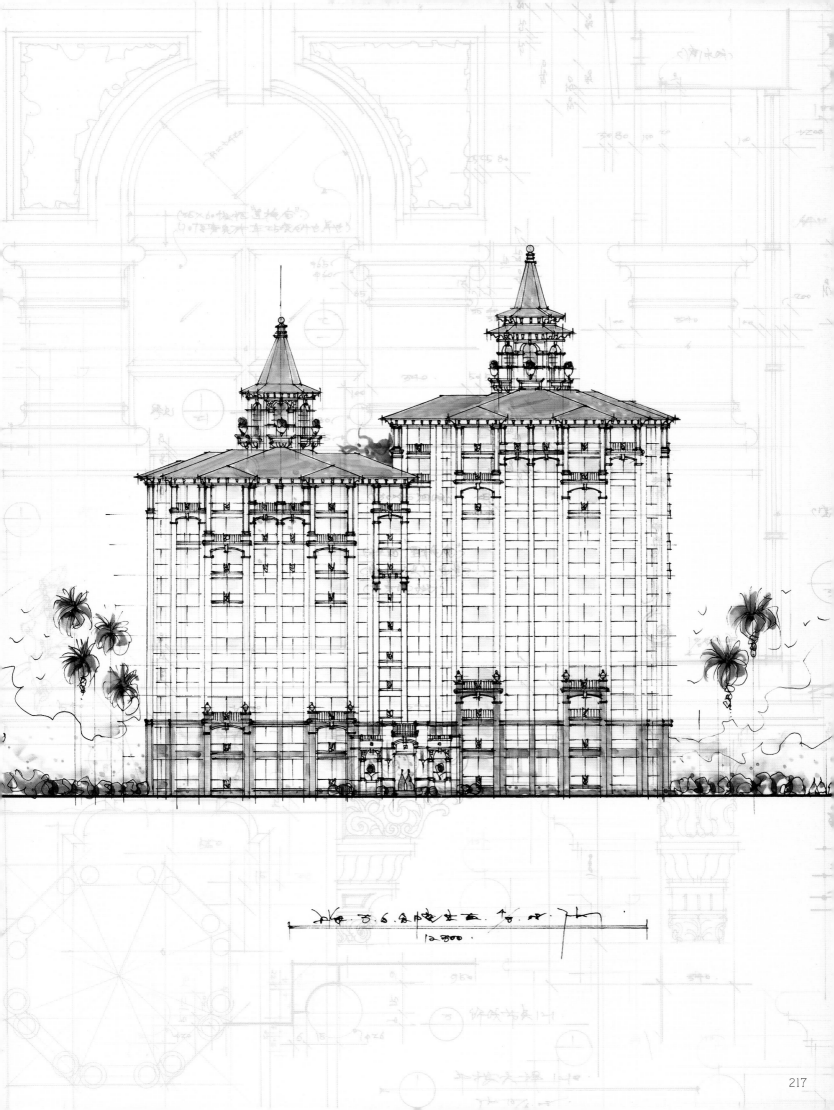

学而知不足

掌声四起，赞誉不绝，

从起步至成功，一步一个脚印，

虽然没有血与泪，但辛勤的汗水，凝聚着的艰辛与苦涩，令人回味。

不断的努力，如同爬上了一座山峰，却发现原来站在另一个山的脚下。

让人喜爱的作品，尚未来得及骄傲与炫耀，却惊叹，自愧它的缺憾与不足，

原来成功让人站得更高，

原来成功让人看得更远，

原来成功可以让人自卑，让人沮丧……

Learning without contentment

Applause and compliment are everywhere.
From the beginning to success, step by step,
though there is no tear and blood,
but the sweat combined with hardness and bitter is still in my memories.
It is like climbing mountain,
after continuous efforts only find yourself at the foot of the other mountains.
You're amazed and shamed of the popular works before had
the chance to be proud and happy about it.
Success can help your stand higher and
see farther but at the same time it makes you self-abased and sad…

作 品 回 顾
WORKS REVIEW

室内设计·建筑设计·规划设计
INTERIOR DESIGN　　　ARCHITECTURE　　　PLANNING DESIGN

1991—2001年

◆

＊祈福新村

连排别墅、独幢别墅、多层公寓

◆

住宅板房、祈福华厦、祈福会所

◆

祈福酒店、南沙酒店、祈福医院

◆

2001年

◆

＊合生创展

华南新城独幢别墅

2002—2013年

◆

＊星河湾集团

广州星河湾

◆

北京星河湾

◆

上海星河湾

◆

星河湾酒店

◆

2006年—至今

◆

＊绿城中国

绿城·上海黄浦湾项目

绿城·温州鹿城广场项目

◆

绿城·绍兴金绿泉项目

◆

绿城·大连深蓝中心项目

绿城·苏州御园项目

◆

绿城·青岛深蓝广场项目

绿城·嘉兴悦庄别墅项目

绿城·唐山南湖春晓项目

绿城·杭州溪上玫瑰园项目

◆

绿城·西溪诚园项目

◆

绿城·舟山玉兰项目

◆

绿城·杭州明月江南项目

◆

绿城·无锡玉兰花园项目

绿城·杭州云栖玫瑰园项目

◆

绿城·常州玉兰广场项目

◆

绿城·杭州玉园项目

绿城·杭州西子御园项目

◆

绿城·上海唐镇项目

绿城·舟山长峙岛项目

◆

绿城·杭州富春和园项目

◆

绿城·无锡蠡湖新城项目

绿城·山东德润御园项目

绿城·湖州御园别墅项目

◆

绿城·慈溪浒山项目

绿城·建德春江明月项目

绿城·金华御园项目

◆

绿城·绍兴新昌兰园项目

绿城·杭州兰园项目

绿城·平阳品致小区项目

◆

绿城·安吉诚园项目

◆

绿城·绍兴玉园项目

◆

绿城·杭州新绿园项目

◆

绿城·杭州赛丽丽园项目

绿城·杭州蓝色钱江项目

◆

绿城·绍兴百合花园项目

◆

绿城·唐山慧翔御园项目

* 滨江集团

杭州滨江湘湖壹号项目

杭州武林壹号项目

◆

* 鲁能仲盛置业

北京鲁能海港格拉斯小镇项目

* **恒力集团**

苏州吴江恒力橡逸湾项目
（规划、建筑、园林、室内设计）

◆

苏州恒力化纤别墅项目

大连恒力维多利亚公馆会所大堂项目

大连恒力维多利亚公馆住宅项目

大连恒力依云伴山项目

◆

苏州恒力滨湖新城9号项目

苏州恒力苏州湾一号项目

◆

* **远洋地产**

北京远洋天著平墅项目

◆

大连远洋自然项目

* 山东鲁信置业

青岛鲁信 长春花园项目

◆

青岛鲁信枣山南海花园项目

青岛鲁信 天逸海湾项目

◆

青岛鲁信随珠花园项目

青岛鲁信和璧花园项目
（规划、建筑、园林、室内设计）

青岛鲁信停云山庄项目

◆

* 华润置地

大连华润星海湾项目

南宁华润幸福里项目

上海华润佘山九里项目

*北京慧诚房地产

温岭湖畔一号项目
（规划、建筑、园林、室内设计）

北京密云君山卧龙山别墅项目

北京密云君山卧龙山庄会所项目

*保利地产

南宁保利君悦湾项目

◆

*华茂置业

宁波华茂·悦峰项目

◆

*荣和集团

南宁荣和大地项目

＊中国葛洲坝集团

海南陵水葛洲坝·海棠福湾会所大堂项目

海南陵水葛洲坝·海棠福湾别墅项目

◆

＊南宁大西洋置业

南宁凯旋一号项目

＊银河湾地产

◆

无锡太湖银河湾项目

◆

常州银河湾第一城项目

无锡银河湾紫苑项目

◆

常州银河湾明苑项目

◆

苏州太仓银河湾花园项目

◆

＊龙光地产

◆

广州南沙龙光棕榈水岸项目

◆

汕头龙光尚海阳光项目

珠海龙光海悦云天项目

* 众彩纷纭

香港鸭脷洲南湾豪宅

香港御龙山豪宅

◆

香港中半山梅道1号豪宅

澳门名门世家项目

◆

深圳康利展厅项目

北京富力城别墅豪宅

◆

北京银信兴业康城花园项目

◆

北京山水文园项目

◆

北京天安国汇项目

◆

北京顺驰百善园豪宅

◆

北京中芯花园项目

厦门中铁元湾豪宅

◆

南昌恒茂天鹅堡豪宅

上海新江湾城银亿领墅项目

◆

广州金穗西苑、东苑、和平大厦项目

◆

广州番禺金海岸项目

佛山敏捷畔海御峰花园项目

◆

江门濠江项目

◆

海南海口蓝城一号项目

海南陵水雅居乐清水湾豪宅

大连颐和星海项目

◆

大连高时集团展厅项目

大连小平岛项目

◆

大连梧桐街8号项目

青岛紫檀山项目

青岛颐和星苑项目

济南金羚·开元府项目

◆

济南绿城御园豪宅

杭州金都高尔夫艺墅项目

杭州桐庐大奇山郡项目

杭州九溪玫瑰园项目

◆

杭州瑞立江河汇项目

◆

杭州昆仑公馆项目

◆

杭州宋都阳光国际三期项目

杭州华润新鸿基之江九里项目

◆

杭州璟熠大家之江悦项目

◆

杭州彩虹城项目

杭州富越香墅排屋项目

温州瑞安天瑞·香山美邸项目

温州中瓯金色海岸项目

◆

温州瑞安天瑞晶品项目

温州瑞安瑞祥新区中房·水岸蓝庭项目

◆

福州泰禾红树林项目

温州瓯江公馆项目

◆

温州瑞安天瑞·尚品项目

福州家天下三木城项目

扬州蜀冈玫瑰园项目

常州巨凝金水岸项目

◆

南京康利华府项目

◆

福州泰禾红峪项目

扬州新能源名门壹品项目

徐州盛和云龙观邸项目

扬州莱茵达瘦西湖·唐郡独栋别墅项目

◆

千岛湖丽湖馨居独立别墅项目

郑州振兴富田兴龙湾项目

成都牧马山三盛·翡俪山项目

合肥栢悦公馆项目

◆

合肥长安天玺项目

◆

合肥同济御园项目

成都中德世纪·英伦联邦项目

◆

沈阳新湖中宝项目

◆

江阳泸州天瑞·尚城项目

哈尔滨四季上东项目
（规划、建筑、园林、室内设计）

重庆融创亚太商谷项目

哈尔滨宏润·翠湖天地项目

媒体《粤商》杂志
陈文栋访谈录

小编：请问几个问题，

>>. 您觉得家居装修最重要的是什么？

答：
最重要当然是：看菜吃饭了。
如果不缺钱，
最重要当然是：舍得。
您自己睡的床垫，品质必须是最好。
如果还没有设计，
最重要当然是：请一个靠谱的设计师……
其实，
家居装修，所有细节都重要！
把细节都做到位，
也就没有最重要的了。

>>. 装修时，怎样选择和搭配艺术品？

答：
我们讲："室内设计"。
"装修"一词，非常不专业！
比如：我是一个建筑师，
你不能说我是一个做土建的。
室内设计：
任何物品的选择与艺术品的陈设、搭配，都没有一个固定的形式。
美学的原理是：
首先要统一，然后讲变化。
在统一的基础上愈变化愈美。
如果是自己的房子，
选择的东西，最重要是自己喜欢。
选择与配搭，讲究色彩与构图。
如果我们不是专业的设计师，
那么就用直接的感受去选择、去配搭。
如同穿衣打扮。反复地推敲，
选、配、自己最舒服的感觉。
喜欢就好！

>>. 怎样确定哪种家居风格适合自己？有时喜欢的未必是适合的。

答：
就是要喜欢！
你喜欢的，就是适合你的。
因为它体现了：
你的品味、你的文化、
你的修养、你的一切。
一目了然，无所遁形。

>>.陈大师对"新古典"这种风格已经驾轻就熟了，啥时候换种风格挑战一下自我？

答：
为什么陈文栋一直在做"新古典"？
难道陈文栋只会做"新古典"？
呵呵！
回答这个问题，要讲故事啦。

2005年，我们请了香港5家有名气的设计公司，设计北京星河湾的交楼标准，但设计没有打动我们。简单如入户大门的设计，也没有丝毫精彩。后来又请了一些国际大师，说实在的，设计很有水平，很有风格，但与"新古典"比，这些设计基本上卖不动。

为什么？
因为市场还没有那么高的欣赏水平，市场也没有文化去理解你的风格。

北京一役、"新古典"风靡全国。

2009年再战上海。
用什么风格来征服这个中国最具自我，最具品味，最具时尚的大都会？在内部我们是经过激烈讨论的。最终还是"新古典"，创造了一天40多个亿的销售纪录。是否"新古典"最受市场欢迎？在崭新的风格成功之前，答案应该是肯定的！

2012年，绿城精装去陈文栋化全面展开。
国际大师、美国的BLD、HBA、法国的PYR粉墨登场，展现他们大师的魅力与国际化的风格。
市场一致认同。
新的风格让高端精装更加丰富多彩，但要知道在这之前，2007—2011年绿城高端精装设计，基本上是本人的作品。
一种风格在一个市场，一定会有它兴衰的过程。
这是市场进步的一种必然。

今年，绿城"上海黄浦湾"，再次让国际大师HBA与陈文栋同台竞技。"新古典"再现舞台。市场的反应，还是"新古典"让人更加喜爱。
（现场销售语）

回到南宁，
2012年，华润"南宁幸福里"邀请我做精装交楼设计。"新古典"再次创造奇迹。幸福里满载而归。

上述举了很多例子，
其实想说明一点：
如果"新古典"是市场的宠儿，
那我们为什么要去颠覆它！

房地产精装交楼，
设计的是："产品"，
不是："作品"。
所有"产品"的创新，
都不是要推翻自己。
"产品"创新，
就是：每天进步一点点！
所有成功的品牌：LV、奔驰、宝马、苹果……
其设计创新，不都是如此而已吗。
展现新的成功的风格，
是一件好事。
但必然有一个过程。
同时也存在风险。
我似乎把风险让给更有力量的国际大师来承担。
绿城，PYR"法式风格"的成功就是一个典范。
其实我现在也在做我的"新法式"。
效果如何，拭目以待。

至于我真正喜欢的风格，
以下的话应代表其心声：
"设计追求时尚，因为时尚是美、
设计必须创新，时尚就是创新。"
你们很多同事都参观了我的新作品，
简约、时尚、大气、奢华的风格，应该是对市场所谓"新古典"的挑战吧。

一个简单的话题谈了这么多。
是因为您关注它。
谈了这么多，
是因为不仅仅是您关注它……

谢谢！
2014／9／1

INTERVIEW
OF CHAN MAN TUNG
BY "CANTONESE BUSINESS" MAGAZINE

Editor: Can I ask you some questions?

>> What is the most important in house decoration?

Mr. Chan: Of course, the most important thing is fitting the appetite to the dishes. If you are not short of money, the key point is not begrudging. The mattress that you use must have high quality. And if it has not been designed, a reliable designer really matters. In fact, for house decoration, every detail is significant. Just try to make all details perfect!

>> For decoration, how to choose and match the artwork?

Mr. Chan: It is "interior design". It is not professional to say "decoration". For example, I am an architect, and it will be improper to say I am the person who works with construction. Interior design: The choice of every article and the layout and match of the artworks are not fixed. The principle of esthetics: Unify first, change second. On the basis of unifying, then the more changing, the more beautiful. If it is you own house, choose what you like. And for selection and matching, pay attention to the color and composition. If we are not professional designers, just choose and match according to you heart and feelings, just like our daily dress. Make repeated scrutiny, selection and matching. You own just like it!

>> How to know which style suits us? You know, sometimes which you likey not fit you.

Mr. Chan: Just need to like it! Which you like must fit you. Because it show your taste, your culture and everything about you! It is rather clear!

>> Professor Chan, you are able to handle "new classical style" with ease. When will you change your style to challenge yourself?

Mr. Chan: Why Chan mantung is always making the "new classical style"? Can he only do this? I must need to tell you a story to answer this question.

In 2005, we invited five well-known design companies from Hong Kong to design the building standards for Beijing Star river. However, that design failed to impress us. Even a simple design for the entrance door was not wonderful at all. Later, we invited a few international professors, to be honest, whose designs are wonderful and stylish. However, such designs cannot be sold compared with "New Classical style".

Why? Because the market has not met the level to appreciate it, and the market also has no culture to understand your style.

After the design in Beijing, "new classical style" swept the country. In 2009, it came to Shanghai. What style can conquer China, which is with ego, taste and fashion?
In the interior, after hot discussion, it is eventually "new classical style". This style creates the record that the sale is more than 4 billion. Is "the new classical style" most popular in the market? Before the success of new style, the answer should be yes!

In 2012, the activity that Greentown try to abandon the style of Chan man tung, was carried overall. The international professor, BLD, HBA from USA and the PYR from France, are on stage, showing their charm and the international style. The market welcomes the new style. The new style indeed makes the advanced decoration more colorful. But you know, the advanced decoration in Greentown is almost my work.

A style in a market must have the process of flourishing and declining. It is inevitable in market's improvement..

This year, for Shanghai Greentown Huangpu bay, the international mater HBA and Chan mantung work together. The "new classical style" comes to the stage again. According to the market reflection, the "new classical style" is more welcomed. (The sales words)

Now back to Nanning,In 2012, Hua Rui "Nanning Xingfuli" invited me to be the designer. The "new classical style" makes wonders again. "Xingfuli" gained a lot.

The many examples above-mentioned in fact prove that if the "new classical style" is welcomed to the market, why we should to overturn it?

What the real estate advanced building design is "the product", not the "work". Every innovation of "product" is not to overturn their own. Instead, make a little progress every day! Just like all the successful brands: LV, Mercedes, Benz, BMW, Apple...
Showing the new successful style is a good thing, but there must be a process and meanwhile, there is a risk. I seem to let the international master who is more powerful than me to take the risk. The success of Greentown:PYR "French style" is an example. In fact, now, I am doing my "new French style". Just wait and see.

For the style I really like, the following words may represent the voice: "design pursue fashion, for fashion is beauty and design need innovation, the fashion is innovation".Many of you have visited my new work, the style which is simple, fashionable, luxury, is maybe a challenge to the so-called "new classical style".We talked a lot about this simple topic, because both you and us pay a lot of attention on it.

Thank you!
Sept.1, 2014

心 声

时光匆匆一去不返,唯有影像让它停留。

刻画出美好的瞬间,雕琢着魅力与光彩。

设计将美丽给别人,生活把友情留自己。

即使是短暂的快乐,回忆却可以是永远!

◆

VOICE

Time blows without turning back, Only in image can it be held, The shining moments it depicts, The glamour and color it sculptures, Design passes beauty to others, Life keeps friendship to ourselves, Flashy happiness it may be, The memory can stays forever!

编 后 语

思念，是一种很玄的东西，如影随形……

从1991年进入中国房地产，成为一名职业经理人、设计师。

到2013年退出这个让人有机会折腾的舞台。

22年弹指一挥间。

从厚积薄发，到鱼跃龙门，至如鱼得水，

品尝了人生的酸、甜、苦、辣……

许多的幸运，许多的知遇，

许多的欣赏，许多的成功，

你经历了，强烈的对比让人心寒，也让人豁达。

也只有如此，

你才会由衷地感谢与珍惜你遇到的人，你获得的机会，你曾经并肩奋斗的良朋益友。

这一切

无声又无息出没在心底……

ADDITIONAL WORDS

Missing is a very mysterious thing lurking as shadow

In 1991, I entered into the industry of Chinese real estate and became a professional manager and designer.

Till 2013, I removed myself from this stage full of opportunity.

22 years passed in the blink of an eye.

From accumulation to success and being professional

Taste all kinds of flavors: sour, sweet, bitter, spicy...

Lots of luck and friends, appreciation and success.

Everything you have experienced chills my heart but makes me generous as well.

Only then, your will gratitude and cherish everyone you have met and every opportunity you've got.

All this silently signed in the bottom of my heart.

香港陈文栋设计师有限公司
CWD designers (hk) Ltd.

◆ http://www.cwd.com.hk ◆

香港陳文棟設計師有限公司
CWD designers (hk) Ltd.

◆ http://www.cwd.com.hk ◆